David Attenborough is one of the world's greatest broadcasters and naturalists, and his career has spanned sixty years, and his extraordinary contribution to natural history broadcasting and film-making has brought him international recognition, from *Life on Earth* to *Frozen Planet*, *Planet Earth* to *Blue Planet*. He has achieved many professional awards, honours and merits, including the CBE and OM, and he was knighted in 1985.

Praise for *Life on Earth* (2018)

'The new *Life on Earth* is as glorious as the first, if not more so for the sole reason that it has been considerably updated'
Adam Rutherford, *Guardian*

'A beautiful and wide ranging work. The breadth of natural history covered is extraordinary and mesmerising. Life on Earth is still breathtakingly rich, and we would know far less about it were it not for Attenborough's wonderful skills of communication over the years: our cultural and scientific lives would be poorer without him'
New Scientist

The Greatest Story Ever Told

Life on Earth
David Attenborough

WILLIAM
COLLINS

William Collins
An imprint of HarperCollins*Publishers*
1 London Bridge Street
London SE1 9GF

WilliamCollinsBooks.com

First published in Great Britain by William Collins Sons & Co. Ltd.
and BBC Books: a division of BBC Enterprises Ltd. in 1979
Fully updated and republished by William Collins in 2018

This William Collins paperback edition published in 2019

2022 2021 2020 2019
10 9 8 7 6 5 4 3 2 1

ISBN 978-0-00-829430-4

Designed and typeset in Bask
Printed and bound in Great I

MIX
Paper from
responsible sources
FSC™ C007454
www.fsc.org

This book is produced from independently cen
paper to ensure responsible forest management
For more information visit: www.harpercollins.

CONTENTS

PROLOGUE

I still recall, with great clarity, the very first time I went to the tropics. Stepping out of the plane and into the muggy, perfumed air of West Africa was like walking into a steam laundry. Moisture hung in the atmosphere so heavily that my skin and shirt were soaked within minutes. A hedge of hibiscus bordered the airport buildings. Sunbirds, glittering with green and blue iridescence, played around it, darting from one scarlet blossom to another, hanging on beating wings as they probed for nectar. Only after I had watched them for some time did I notice, clasping a branch within the hedge, a chameleon, motionless except for its goggling eyes, which swivelled to follow every passing insect. Beside the hedge, I trod on what appeared to be grass. To my astonishment, the leaflets immediately folded themselves flat against the stem, transforming green fronds into apparently bare twigs. It was sensitive mimosa. Beyond lay a ditch covered with floating plants. In the spaces between them, the black water rippled with fish, and over the leaves walked a chestnut-coloured bird, lifting its long-toed feet with the exaggerated care of a man in snowshoes. Wherever I looked, I found a prodigality of pattern and colour for which I was quite unprepared. It was a revelation of the splendour and fecundity of the natural world from which I have never recovered.

Since then, I have managed, one way or another, to get back to the tropics many times. Usually my purpose has been to make a film about some corner of that infinitely varied world. So I have had the luck to find and film rare creatures that few outsiders have seen in the wild, and to gaze on some of the most marvellous spectacles that the wild places of the world have to offer – a tree full of displaying birds of paradise in New Guinea, giant lemurs leaping through the forest of Madagascar, the biggest lizards in the world prowling, like dragons, through the jungle of a tiny island in Indonesia.

Initially, the films we made tried to document the lives of particular animals showing how each found its food, defended itself and courted, and the ways in which it fitted into the community of animals and plants around it. But then the idea formed in my mind that a group of us might make a series of films that portrayed animals in a slightly different way. Our subject would be not only natural history in the sense that those two words are normally used, but the history of nature. We would try to survey the whole animal kingdom and consider each great group of animals in the light of the part it has played in the long drama of life from its beginnings until today. This book originated from the three years of travelling and research that went into the making of those films.

The condensation of three thousand million years of history into three hundred or so pages, and the description of a group of animals containing tens of thousands of species within one chapter, compels vast omissions. My method was to try to perceive the single most significant thread in the history of a group and then concentrate on tracing that, resolutely ignoring other issues, no matter how enticing they may seem.

This, however, risks imposing an appearance of purpose on the animal kingdom that does not exist in reality. Darwin demonstrated that the driving force of evolution comes from the accumulation, over countless generations, of chance genetic changes sifted by the rigours of natural selection. In describing the consequences of this process it is only too easy to use a form of words that suggests that the animals themselves were striving to bring about change in a purposeful way – that fish *wanted* to climb on to dry land and to modify their fins into legs, that reptiles *wished* to fly, strove to change their scales into feathers and so ultimately became birds. There is no objective evidence of anything of the kind and I have endeavoured, while describing these processes in a reasonably succinct way, not to use any phrases that might suggest otherwise.

To a surprising degree, nearly all the major events in this history can be told using living animals to represent the ancestral creatures which were the actual protagonists.

The lungfish today shows how lungs may have developed; the mouse deer represents the first hoofed mammals that browsed in the forests of fifty million years ago. But misunderstandings can come unless the nature of this impersonation is made quite clear. In rare instances, a living species seems to be identical with one whose remains are fossilised in rocks several hundred million years old. It happens to have occupied a niche in the environment that has existed unchanged for such vast periods of time and suited it so ideally that it had no cause to change. In most cases, however, living species, while they may share essential characters with their ancestors, differ from them in many ways. The lungfish and the mouse deer are fundamentally similar to their ancestors, but they are by no means identical. To underline this distinction each time with a phrase like 'ancestral forms that closely resemble the living species' would be unnecessarily clumsy and literal-minded, but that qualifying phrase must be taken as read whenever I have referred to an ancient creature by the name of a living one.

Since this book was first written, science of course has continued to make new discoveries that have illuminated and amplified the history of nature. New species – some living, some fossil – have been discovered that link different groups. Some discoveries have been truly sensational. Perhaps the most dramatic have been those made in China of small dinosaurs with the clearly identifiable remains of feathers covering many parts of their bodies. They have cleared up one of the great and most vehement arguments among evolutionary biologists about the origins of both flight and the birds. Another concerns the very origins of life itself. Fossils have been found not only in Australia but in many other parts of the world, including the Avalon peninsula in northern Canada where a seabed thronged with all kinds of hitherto unknown organisms and dating from around 565 million years ago has been preserved with astounding perfection. All these advances in knowledge and many more will be mentioned in the appropriate places in the text that follows.

One whole new branch of science has in recent years spread a great deal of light on the history of life – molecular genetics. Nearly a century after the publication of Darwin's book on evolution by natural selection, *On the Origin of Species*, Crick and Watson described the structure of deoxyribonucleic acid – DNA for short – the molecule that carries the genetic blueprint from which another individual animal can be developed. This explained the mechanism by which physical characteristics are passed from one generation to the next.

The first organism to have its version completely deciphered was a small worm. Once that was done, the next great target was to analyse human DNA. That took many years of both international competition and cooperation. Today, however, it is possible to establish genetic identity of a species in a few hours using a piece of apparatus no bigger than a mobile phone. With such knowledge and techniques. all kinds of things can now be deduced – the relationship between individual species, the date in its evolutionary history at which any particular characteristic appeared, and even the precise way in which it did so. So the connections between the various groups that appear in our story can now be determined and statements about ancestry made with real confidence. Such new insights will be described in this new edition in their appropriate places in the pages that follow.

I have used familiar English names rather than scientific Latin ones so that when an animal makes its appearance in this history, it is quickly recognised for what it is. Those who wish to discover more about it in more technical books will find its scientific name in the index. For the most part, I have expressed age in absolute terms of millions of years rather than use the adjectival names of periods coined by classical geology. Lastly, I have made no reference by name to those many scientists whose work has provided the facts and theories on which the following pages are based. This has been done solely to try to maintain clarity in the narrative. I intend no minimisation of the debt owed to them by all of us who take pleasure in watching and thinking about animals. They and their researches have provided us with that most valuable of insights, the ability to perceive the continuity of nature in all its manifestations and to recognise our place within it.

ONE

The Infinite Variety

It is not difficult to discover an unknown animal. Spend a day in the tropical forest of South America, turning over logs, looking beneath bark, sifting through the moist litter of leaves, followed by an evening shining a mercury lamp on a white screen, and one way or another you will collect hundreds of different kinds of small creatures. Moths, caterpillars, spiders, long-nosed bugs, luminous beetles, harmless butterflies disguised as wasps, wasps shaped like ants, sticks that walk, leaves that open wings and fly – the variety will be enormous and one of these creatures is quite likely to be undescribed by science. The difficulty will be to find specialists who know enough about the groups concerned to be able to single out the new one.

No one can say just how many species of animals there are in these greenhouse-humid dimly lit jungles.

They contain the richest and the most varied assemblage of animals and plant life to be found anywhere on earth. Not only are there many major categories of creatures – monkeys, rodents, spiders, hummingbirds, butterflies – but most of those types exist in many different forms. There are over forty different species of parrot, over seventy different monkeys, three hundred hummingbirds and tens of thousands of butterflies. If you are not careful, you can even be bitten by a hundred different kinds of mosquito.

In 1832 a young Englishman, Charles Darwin, twenty-four years old and naturalist on HMS *Beagle*, a brig sent by the Admiralty in London on a surveying voyage round the world, came to such a forest outside Rio de Janeiro. In one day, in one small area, he collected sixty-eight different species of small beetle. That there should be such a variety of species of one kind of creature astounded him. He had not been searching specially for them so that, as he wrote in his journal, 'It is sufficient to disturb the composure of an entomologist's mind to look forward to the future dimensions of a complete catalogue'. The conventional view of his time was that all species were immutable and that each had been individually and separately created by God. At the time, Darwin was far from being an atheist – he had, after all, taken a degree in divinity at Cambridge University – but he was deeply puzzled by this enormous multiplicity of forms.

During the next three years, the *Beagle* sailed down the east coast of South America, rounded Cape Horn and came north again up the coast of Chile. The expedition then sailed out into the Pacific until, 1,000 kilometres from the mainland, they came to the lonely archipelago of the Galapagos. Here Darwin's questions about the creation of species recurred, for in these islands he found fresh variety. He was fascinated to discover that the Galapagos animals bore a general resemblance to those he had seen on the mainland, but differed from them in detail. There were cormorants, black, long-necked diving birds like those that fly low along Brazilian rivers, but here in the Galapagos, their wings were so small and with such stunted feathers that they had lost the power of flight. There were iguanas, large lizards with a crest of scales along their backs. Those on the continent climbed trees and ate leaves. Here on the islands, where there was little vegetation, one species fed on seaweed and clung to rocks among the surging waves with unusually long and powerful claws. There were tortoises, very similar to the mainland forms except that these were many times bigger, giants that a man could ride. The British Vice-Governor of the Galapagos told Darwin that even within the archipelago, there was variety: the tortoises on each island were slightly different, so that it was possible to tell which island they came from. Those that lived on relatively well watered islands where there was ground vegetation to be cropped, had a gently curving front edge to their shells just above the neck. But those that came from arid islands and had to crane their necks in order to reach branches of cactus or leaves of trees, had much longer necks and a high peak to the front of their shells that enabled them to stretch their necks almost vertically upwards.

slightly longer necks than others. In times of drought they will be able to reach leaves and so survive. Their brothers and sisters, with shorter necks, will starve and die. So those best fitted to their surroundings will be selected and be able to transmit their characteristics to their offspring. After a great number of generations, tortoises on the arid islands will have longer necks than those on the watered islands – one species will have given rise to another.

This concept did not become clear in Darwin's mind until long after he had left the Galapagos. For twenty-five years he painstakingly amassed evidence to support it. Not until 1859, when he was forty-eight years old, did he publish it, and even then he was driven to do so only because another younger naturalist, Alfred Wallace, working in Southeast Asia, had formulated the same idea. He called the book in which he set out his theory in detail, *On the Origin of Species by Means of Natural Selection, or the Preservation of Favoured Races in the Struggle for Life.*

Since that time, the theory of natural selection has been debated and tested, refined, qualified and elaborated. Later discoveries about genetics, molecular biology, population dynamics and behaviour have given it new dimensions. It remains the key to our understanding of the natural world and it enables us to recognise that life has a long and continuous history during which organisms, both plant and animal, have changed, generation by generation, as they colonised all parts of the world.

There are now two direct sources of evidence for this history. One can be found in the genetic material in the cells of every living organism. The other lies in the archives of the earth, the sedimentary rocks. The vast majority of animals leave no trace of their existence after their passing. Their flesh decays, their shells and their bones become scattered and turn to powder. But very occasionally, one or two individuals out of a population of many thousands have a different fate. A reptile becomes stuck in a swamp and dies. Its body rots but its bones settle into the mud. Dead vegetation drifts to the bottom and covers them. As the centuries pass and more vegetation accumulates, the deposit turns to peat. Changes in sea level may cause the swamp to be flooded and layers of sand to be deposited on top of the peat. Over great periods of time, the peat is compressed and turned to coal. The reptile's bones still remain within it. The great pressure of the overlying sediments and the mineral-rich solutions that circulate through them cause chemical changes in the calcium phosphate of the bones. Eventually they are turned to stone, but they retain the outward shape that they had in life, albeit sometimes distorted. On occasion, even their detailed cellular structure is preserved so that you can look at sections of them through the microscope and plot the shape of the blood vessels and the nerves that once surrounded them. In rare cases, even the colour of skin or feathers can be detected.

The most suitable places for fossilisation are in seas and lakes where sedimentary deposits that will become sandstones and limestones are slowly accumulating. On land, where for the most part rocks are not built up by deposition but broken down by erosion, deposits such as sand dunes are only very rarely created and preserved. In consequence, the only land-living organisms likely to be fossilised are those that happen to fall into water. Since this is an exceptional fate for most of them, we are never likely to know from fossil evidence anything approaching the complete range of land-living animals and plants that has existed in the past. Water-living animals, such as fish, molluscs, sea urchins and corals, are much more promising candidates for preservation. Even so, very few of these perished in the exact physical and chemical conditions necessary for fossilisation. Of those that did, only a tiny proportion happen to lie in the rocks that outcrop on the surface of the ground today; and of these few, most will be eroded away and destroyed before they are discovered by fossil hunters. The astonishment is that, in the face of these adverse odds, the fossils that have been collected are so numerous and the record they provide so detailed and coherent.

How can we date them? Since the discovery of radioactivity scientists have realised that rocks have a geological clock within them. Several chemical elements decay with age, producing radioactivity in the process. Potassium turns into argon, uranium into lead, rubidium into strontium. The rate at which this happens can be estimated. So if the proportion of the secondary element to the primary one in a rock is measured, the time at which the original mineral was formed can be calculated. Since there are several such pairs of elements decaying at different speeds, it is possible to make cross-checks.

This technique, which requires extremely sophisticated methods of analysis, will always remain the province of the specialist. But anyone can date many rocks in a relative way by simple logic. If rocks lie in layers, and are not grossly disturbed, then the lower layer must be older than the upper. So we can follow the history of life through the strata and trace the lineages of animals back to their beginnings by going deeper and deeper into the earth's crust.

The deepest cleft that exists in the earth's surface is the Grand Canyon in the western United States. The rocks through which the Colorado River has cut its way still lie roughly horizontally, layer upon layer, red, brown and yellow, sometimes pink in early light, sometimes blue in the shadowed distance. The land is so dry that only isolated juniper trees and low scrub freckle the surface of the cliffs, and the rock strata, some soft, some hard, are clear and stark. Most of them are sandstones or limestones that were laid down at the bottom of the shallow seas that once covered this part of North America. When they are examined closely, breaks in the succession can be detected. These represent times when the land rose, the seas drained away and the seabed became dry so that the deposits that had accumulated on it were eroded away. Subsequently, the land sank again, seas flooded back and deposition restarted. In spite of these gaps, the broad lines of the fossil story remain clear.

A mule will carry you in an easy day's ride from the rim to the very bottom of the Canyon. The first rocks you pass are already some 200 million years old. There are no remains of mammals or birds in them, but there are traces of reptiles. Close by the side of the trail, you can see a line of tracks crossing the face of a sandstone boulder. They were made by a small four-footed creature, almost certainly a lizard-like reptile, running across a beach. Other rocks, at the same level elsewhere, contain impressions of fern leaves and the wings of insects.

Halfway down the Canyon, you come to 400-million-year-old limestones. There are no signs of reptiles to be found here, but there are the bones of strange armoured fish. An hour or so later – and a hundred million years earlier – the rocks contain no sign of backboned animals of any kind. There are a few shells and worms that have left behind a tracery of trails in what was the muddy seafloor. Three-quarters of the way down, you are still descending through layers of limestone, but now there is no sign of fossilised life whatever. By the late afternoon, you ride at last into the lower gorge where the Colorado River runs green between high rock walls. You are now well over a vertical kilometre below the rim, and the surrounding rocks have been dated to the immense age of 2,000 million years. Here you might hope to find evidence for the very beginnings of life. But there are no organic remains of any kind. The dark fine-grained rocks lie not in horizontal layers like all those above, but are twisted and buckled and riven with veins of pink granite.

Are signs of life absent because these rocks and the limestones directly above are so extremely ancient that all such traces have been crushed from them? Could it be that the first creatures to leave any sign of their existence were as complex as worms and molluscs? For many years these questions puzzled geologists. All over the world, rocks of this antiquity were carefully searched for organic remains. One or two odd shapes were found, but most authorities dismissed these as patterns produced by the physical processes of rock formation that had nothing whatever to do with living organisms. Then during the 1950s, the searchers began to use high-powered microscopes on some particularly enigmatic rocks.

Around 1,600 kilometres northeast of the Grand Canyon, ancient rocks of about the same age as those beside the Colorado River outcrop on the shores of Lake Superior. Some of them contain seams of a fine-grained flint-like substance called chert. This was well known during the nineteenth century because the pioneers used it in their flintlock guns. Here and there, it contains strange white concentric rings a metre or so across. Were these merely eddies in the mud on the bottom of the primeval seas, or could they have been formed by living organisms? No one could be sure and the shapes were given the noncommittal name of stromatolite, a word derived from Greek meaning no more than 'stony carpet'. But when researchers cut sections of these rings, ground them down into slices so thin that they were translucent and examined them through the microscope, they found, preserved in the chert, the shapes of simple organisms, each no more than one or two hundredths of a millimetre across. Some resembled filaments of algae; others, while they were unmistakably organic, had no parallels with living organisms; and some looked to be identical with the simplest form of life existing today: bacteria.

It seemed almost impossible to many people that such tiny things as microorganisms could have been fossilised at all. That relics of them should have survived for such a vast period of time seemed even more difficult to believe. The solution of silica which had saturated the dead organisms and solidified into chert was clearly as fine-grained and durable a preservative as exists. The discovery of the fossils in the Gunflint Chert stimulated further searches not only in North America but all over the world, and other microfossils were found in cherts in Africa and Australia. Some of these, astonishingly, pre-dated the Gunflint specimens by a billion years, and some scientists now claim to have found fossils from around 4 billion years ago, not long after the formation of the earth. But if we want to consider how life arose, fossils cannot help us, for the origin of life involved the interaction of molecules, which leave no fossil traces. To understand what scientists think happened we have to look back beyond even the earliest microfossils, to a time when the earth was completely lifeless.

In many ways the planet then was radically different from the one we live on today. There were seas, but the way the land masses lay bore no resemblance in either form or distribution to modern continents. Volcanoes were abundant, spewing noxious gases, ash and lava. The atmosphere consisted of swirling clouds of hydrogen, carbon monoxide, ammonia and methane. There was little or no oxygen. This unbreathable mixture allowed ultraviolet rays from the sun to bathe the earth's surface with an intensity that would be lethal to modern animal life. Electrical storms raged in the clouds, bombarding the land and the sea with lightning.

Laboratory experiments were made in the 1950s to discover what might happen to these particular chemical constituents under such conditions. Such gases, mixed with water vapour, were subjected to electrical discharge and ultraviolet light. After only a week of this treatment complex molecules were found to have formed in the mixture, including sugars, nucleic acids and amino acids, the building blocks of proteins. We now know that such simple organic molecules can be found throughout the universe, including on interstellar bodies such as comets. But amino acids are not life, nor are they even necessary for life to exist. The experiment proved little about the origin of life.

All forms of life that exist today share a common way of transmitting genetic information, of telling cells what to do. It is a molecule called deoxyribonucleic acid, or DNA for short. Its structure gives it two key properties. First, it can act as a blueprint for the manufacture of amino acids; and second, it has the ability to replicate itself. With this substance, molecules had reached the threshold of something quite new. These two characteristics of DNA also characterise even the simplest of living organisms such as bacteria. And bacteria, besides being the simplest form of life we know, are also among the oldest fossils we have discovered.

The ability of DNA to replicate itself is a consequence of its unique structure. It is shaped like two intertwined helices. During cell division, these unzip, splitting the molecule along its length into two separate helices. Each then acts as a template to which other simpler molecules become attached until each has once more become a double helix.

The simple molecules from which the DNA is mainly built are of only four kinds, but they are grouped in trios and arranged in a particular and significant order on the immensely long DNA molecule. This order specifies how the twenty or so different amino acids are arranged in a protein, how much is to be made, in what tissue and when. A length of DNA bearing such information for a protein, or for how a protein should be expressed, is called a gene.

Occasionally, the DNA copying process involved in reproduction may go wrong. A mistake may be made at a single point, or a length of DNA may become temporarily dislocated and be reinserted in the wrong place. The copy is then imperfect and the proteins it will create may be entirely different. Changes in the DNA sequence can also be induced by chemicals or radiation. When this occurred in the first organisms on earth, evolution began, for such hereditary changes, brought about by mutation and errors, are the source of variations from which natural selection can produce evolutionary change.

Because all life shares DNA as the hereditary material, it is possible to compare DNA sequences in different organisms and show how they are related. Such is the progress of technology that it is now also possible to sequence all the DNA in an organism in a matter of hours, using a device the size of a mobile phone. The millions of DNA sequences that have been established, stored in databases and compared show us unequivocally that, just as Darwin predicted, all life on earth shares a common ancestor. Because parts of our DNA accumulate mutations at a constant rate, like a molecular clock, we can use DNA sequences to estimate when two species split apart. In general, genetic and fossil timings agree with each other, although genetic data do sometimes throw up surprises. Using this method we can estimate that the Last Universal Common Ancestor of all life on earth – commonly known as LUCA, and basically a population of simple bacteria – lived around 4 billion years ago. Everything we can see around us can trace its ancestry back to that group of cells.

Such vast periods of time baffle the imagination, but we can form some idea of the relative duration of the major phases of the history of life if we compare the entire span, from these first beginnings until today, with one year. That means that, roughly, each day represents around ten million years. On such a calendar, the Gunflint fossils of algae-like organisms, which seemed so extremely ancient when they were first discovered, are seen to be quite late-comers in the history of life, not appearing until the second week of August. In the Grand Canyon, the oldest worm trails were burrowed through the mud in the second week of November and the first fish appeared in the limestone seas a week later. The little lizard will have scuttled across the beach during the middle of December and humans did not appear until the evening of 31 December.

But we must return to January. The bacteria fed initially on the various carbon compounds that had taken so many millions of years to accumulate in the primordial seas, producing methane as a by-product. Similar bacteria still exist today, all over the planet. And that was all there was, for around five or six months of our year. Then, in the early summer of the year of life, so some time over 2 billion years ago, bacteria developed an amazing biochemical trick. Instead of taking ready-made food from their surroundings, they began to manufacture their own within their cell walls, drawing the energy needed to do so from the sun. This process is called photosynthesis. One of the ingredients required by the earliest form of photosynthesis is hydrogen, a gas that is produced in great quantities during volcanic eruptions.

Conditions very similar to those in which the early photosynthesising bacteria lived can be found today in such volcanic areas as Yellowstone in Wyoming. Here a great mass of molten rock, lying only a thousand metres or so, down in the earth's crust, heats the rocks on the surface. In places, the ground water is well above boiling point. It rises up channels through the rocks under decreasing pressure until suddenly it flashes into steam and water spouts high into the air as a geyser. Elsewhere, the water wells up into steaming pools. As it trickles away and cools, the salts it gathered from the rocks on its way up, together with those derived from the molten mass far below, are deposited to form rimmed and buttressed basins, surrounded by tiers of terraces. In these scalding mineral-laden waters, bacteria flourish. Some grow into matted filaments and curds, others into thick leathery sheets. Many are brilliantly coloured, their intensity of hue varying during the year as the colonies wax and wane. The names given to these pools hint at the variety of the bacteria and the splendour of the effects they produce – Emerald Pool, Sulphur Cauldron, Beryl Spring, Firehole Falls, Morning Glory Pool and – a particularly rich one with several species of bacteria – Artists' Paintpots.

When you wander through this amazing landscape, you can smell sulphurated hydrogen, the unmistakable stench of rotting eggs, produced by the reaction of ground water with the molten rock far beneath. This is the source from which many of the bacteria here obtain their hydrogen, and as long as bacteria were dependent upon volcanic action for it, they could not spread widely. But other forms eventually arose which were able to extract hydrogen from a very much more widespread source – water. This development was to have a profound effect on all life to come, for if hydrogen is extracted from water, the element that remains is oxygen. The organisms that did this are barely more complex in structure than bacteria. They are sometimes called blue-green algae because they appeared to be close relatives of the green algae that are common in ponds, but now we realise they are similar to the ancestors of those algae, and they are referred to as cyanobacteria or, simply, blue-greens. The chemical agent which they contain, making it possible for them to use water in the photosynthetic process, is chlorophyll, which is also possessed by true algae and plants.

Blue-greens are found wherever there is constant moisture. You can often see mats of them, beaded with silver bubbles of oxygen, blanketing the bottoms of ponds. In Shark Bay, on the northwest coast of tropical Australia, they have developed in a particularly spectacular and significant form. Hamelin Pool, one small arm of this vast inlet, has its entrance blocked by a sand bar covered with eel grass. The flow of water in and out of the Pool is so greatly impeded that evaporation under the grilling sun has made the waters very salty indeed. As a result, marine creatures such as molluscs which would normally feed on blue-greens and keep them in check, cannot survive. The blue-greens, therefore, flourish uncropped just as they did when they were the most advanced form of life anywhere in the world. They secrete lime, forming stony cushions near the shores of the Pool and teetering columns at greater depths. Here is the explanation of those mysterious shapes seen in section in the Gunflint Chert. The blue-green pillars of Hamelin Pool are living stromatolites, and the groups of them standing on the sun-dappled seafloor are as close as we may ever get to a scene from the world of 2 billion years ago.

The arrival of the blue-greens marked a point of no return in the history of life. In ways we do not fully understand, the oxygen they produced eventually accumulated over the millennia to form the kind of oxygen-rich atmosphere that we know today. Our lives, and those of all other animals, depend on it. We need it not only to breathe but to protect us. Oxygen in the atmosphere forms a screen, the ozone layer, which cuts off most of the ultraviolet rays of the sun.

Life remained at this stage of development for a vast period. Then, around 2 billion years ago one single-celled life form found itself trapped inside another, in an entirely chance encounter. You can find examples of the kind of organisms it eventually produced in almost any patch of fresh water.

A drop from a pond, viewed through a microscope, swarms with tiny organisms, some spinning, some crawling, some whizzing across the field of vision like rockets. As a group they are often called the protozoa, or protists – the name means 'first animals', although they are now seen as a very disparate group, not all of which have any affinity with animals. They are all single cells, yet within their cell walls they contain much more complex structures than any bacterium possesses. One central packet, the nucleus, is full of DNA. This appears to be the organising force of the cell. Elongated bodies, the mitochondria, provide energy by burning oxygen in much the same way as many bacteria do. Many cells have a thrashing tail attached to them and this resembles a filamentous bacterium called a spirochete. Some also contain chloroplasts, packets of chlorophyll which, like blue-greens, use the energy of sunlight to assemble complex molecules as food for the cell. Each of these tiny organisms thus appears to be a committee of simpler ones. This, in effect, is what they are. The mitochondria are the descendants of the single-celled organism that was trapped some 2 billion years ago, say in June in the year of life, while the chloroplasts are descended from a trapped blue-green.

Protozoans reproduce by splitting into two, as bacteria do, but their internal structure is much more complex and their division, not surprisingly, is consequently an elaborate business. Most of the separate structures, the members of the committee, themselves split. Indeed, the mitochondria and chloroplasts, each with their own DNA as befits their origins as separate organisms, often do so independently of the division of the main cell. The DNA within the nucleus replicates in a particularly complex manner which ensures that all its genes are copied and that each daughter cell receives a complete duplicate set. There are, however, several other methods of reproduction practised by various protozoans on occasions. The details vary. The essential feature of all the techniques is that a shuffling of genes is involved. In some cases this takes place when two cells join up and exchange genes before breaking apart and then undergoing cell division some time later. In other cases, cells normally contain two complete sets of genes which, after shuffling, divide to make new cells with only one set. These cells are of two types – a large comparatively immobile one, and a smaller active one, driven by a flagellum. The first is called an egg and the second a sperm – for this is the dawning of sex. When the two types unite in a new amalgamated cell the genes are once again in two sets but in new combinations with genes from not just one parent but two. This may well be a unique combination which will produce a slightly different organism with new characteristics. Since the evolution of sex increased the possibilities of genetic variation, it also greatly accelerated the rate at which evolution could proceed as organisms encountered new environments.

There are tens of thousands of species of protozoans. Some are covered by a coat of flailing threads or cilia, which with a coordinated beat drive the creature through the water. Others, including the amoeba, move by bulging out fingers from the main body and then flowing into them. Many of those that live in the sea secrete shells with the most elaborate structure of silica or calcium carbonate. These are among the most exquisite objects that the microscope-carrying explorer will ever encounter. Some resemble minuscule snail shells, some ornate vases and bottles. The most delicate of all are of shining translucent silica, concentric spheres transfixed by needles, gothic helmets, rococo belfries and spiked space capsules. The inhabitants of these shells extend long threads through pores with which they trap particles of food.

Other protists feed in a different way, photosynthesising with the aid of their packets of chlorophyll. These can be regarded as plants; the remainder of the group, which feed on them, as animals. The distinction between the two at this level, however, does not have as much meaning as such labelling might suggest, for there are many species that can use both methods of feeding at different times.

Some protists are just large enough to see with the naked eye. With a little practice, the creeping grey speck of jelly which is an amoeba can be picked out in a drop of pond water. But there is a limit to the growth of a single-celled creature, for as size increases, the chemical processes inside the cell become difficult and inefficient. Size, however, can be achieved in a different way – by grouping cells together in an organised colony.

One species that has done this is volvox, a hollow sphere, almost the size of a pinhead, constructed from a large number of cells, each with a flagellum. The striking thing about these units is that they are virtually the same as other single cells that swim by themselves and have separate existences. The constituent cells of volvox, however, are coordinated, for all the flagella around the sphere beat in an organised way and drive the tiny ball in a particular direction.

This kind of coordination between constituent cells in a colony was taken a stage further, probably between 800 and 1,000 million years ago – some time in October in our calendar – when sponges appeared. Sponges can grow to a very considerable size. Some species form soft shapeless lumps on the seafloor two metres or so across. Their surfaces are covered with tiny pores through which water is drawn into the body by flagella, and then expelled through larger vents. The sponge feeds by filtering particles from this stream of water passing through its body. The colonial bonds between its constituents are very loose. Individual cells may crawl about over the surface of the sponge like amoebae. If two sponges of the same species are growing close to one another, they may, as they grow, come into contact and eventually merge into one huge organism. If a sponge is forced through a fine gauze sieve so that it is broken down into separate cells, these will eventually reorganise themselves into a new sponge, each kind of cell finding its appropriate place within the body. Most remarkably of all, if you take two sponges of the same species and treat them both in this extreme way and then mix cells from the two, they will reconstitute themselves into a single mixed-parentage entity.

Some sponges produce a soft, flexible substance around their cells which supports the whole organism. This, when the cells themselves have been killed by boiling and washed away, is what we use in our baths. Other sponges secrete tiny needles, called spicules, either of calcium carbonate or silica, which mesh together to form a scaffold in which the cells are set. How one cell orientates itself and produces its spicule so that it fits perfectly into the overall design is totally unknown. When you look at a complex sponge skeleton such as that made of silica spicules which is known as Venus' flower basket, the imagination is baffled. How could quasi-independent microscopic cells collaborate to secrete a million glassy splinters and construct such an intricate and beautiful lattice? We do not know. But even though sponges can produce such miraculous complexities as this, they are not like other animals. They have no nervous system, no muscle fibres. The simplest creatures to possess these physical characteristics are the jellyfish and their relatives.

A typical jellyfish is a saucer fringed with stinging tentacles. This form is called a medusa after the unfortunate woman in a Greek myth who was loved by the god of the sea and as a result had her hair changed by a jealous goddess into snakes. Jellyfish are constructed from two layers of cells. The jelly which separates them gives the organism a degree of rigidity needed to withstand the buffeting of the sea. They are quite complex creatures. Their cells, unlike those of the sponge, are incapable of independent survival. Some are modified to transmit electric impulses and are linked into a network which amounts to a primitive nervous system; others are able to contract in length and so can be considered as simple muscles. There are also stinging cells with coiled threads inside them, the unique possessions of the jellyfish tribe. When food or an enemy comes near, the cell discharges the thread, which is armed with spines like a miniature harpoon and often loaded with poison. It is these cells in the tentacles that will sting you if you unluckily brush against a jellyfish when swimming.

Jellyfish reproduce by releasing eggs and sperm into the sea. The fertilised egg does not develop into another jellyfish directly but becomes a free-swimming creature quite different from its parents. It eventually settles down on the bottom of the sea and grows into a tiny flower-like organism called a polyp. In some species, this sprouts, through branching twigs, into other polyps. They filter-feed with the aid of tiny beating cilia. Eventually, the polyps bud in a different way and produce miniature medusae which detach themselves and wriggle away to take up the swimming life once more.

This alternation of form between generations has allowed all kinds of variations within the group. The true jellyfish spend most of their time as free-floating medusae with only the minimum period fixed to the rocks. Others, like the sea anemones, do the reverse. For all their adult lives they are solitary polyps, glued to the rock, their tentacles waving in the water ready to trap prey that may touch them. Yet a third kind are colonies of polyps but ones that have, confusingly, given up their attachment to the sea bottom and sail free like medusae. The Portuguese man o'war is one of these. Chains of polyps dangle from a float filled with gas. Each chain has a specialised function. One kind produces reproductive cells; another absorbs sustenance from captured prey; another, heavily armed with particu-larly virulent stinging cells, trails behind the colony for up to fifty metres, paralysing any fish that blunder into it.

It seems an obvious assumption that these relatively simple organisms appeared very early in the history of animal life, but for a long time there was no proof that they actually did so. Hard evidence could only come from the rocks. Even if microorganisms can be preserved in chert, it is difficult to believe that a creature as large but as fragile and insubstantial as a jellyfish could retain its shape long enough to be fossilised. But in the 1940s some geologists noticed very odd shapes in the ancient Ediacara Sandstones of the Flinders Ranges in southern Australia. These rocks, now thought to be about 650 million years old, were believed to be completely unfossiliferous. Judging from the size of the sand grains of which they are composed and the ripple marks on the surface of their bedding planes, they had once formed a sandy beach. Very occasionally, flower-like impressions were detected on them, some the size of a buttercup, some as big as a rose. Could these be the marks left by jellyfish stranded on the beach, baked in the sun and then covered by a wash of fine sand by the next tide? Eventually enough of these shapes were collected and studied for it to be undeniable that this is just what they must be.

Since then, other assemblages of living organisms of this extreme age have been discovered in many parts of the world – the Charnwood Forest in the heart of England, Namib Desert in southwest Africa, on the flanks of the Ural Mountains and the shores of the White Sea in Russia. But the most impressive and richest of all these discoveries have been made on the Avalon peninsula in Newfoundland. There the rocks, which are around

565 million years old, are exposed in dramatic cliffs. The strata have been tilted and folded, as one might expect in deposits of such extreme age, but not so severely that they have destroyed or even seriously distorted the fossils they contain. These are so abundant that in places it is impossible to walk over the exposed surface of a layer without treading on examples that any museum in the world would regard as one of its greatest treasures. They have been preserved in extraordinary perfection, seemingly by falls of volcanic ash from nearby volcanoes which buried them almost instantaneously, so creating what have been called death masks. There is a rich variety of shapes that are still being catalogued – spindles, fronds, discs, mats, plumes and combs, by far the richest record of any of the communities that flourished in the seas of the world during this extremely ancient period. Many seem to be unrelated to anything alive today and may perhaps be regarded as evolution's failed experiments. One or two, however, bear at least a superficial resemblance to living marine creatures called sea pens that are still common today.

The name sea pen was given them when people wrote with quills, and very apt it must have seemed, for not only are they shaped like feathers but their skeleton is flexible and horny. They grow sticking up vertically on sandy seafloors, some only a few centimetres long, some half as tall as a man. At night they are particularly spectacular for they glow with a bright purple luminescence, and if you touch them, ghostly waves of light pulsate along their slowly writhing arms.

Sea pens are also called soft corals. Stony corals, their relatives, often grow alongside them and they too are colonial creatures. Their history is not as ancient as that of the sea pens, but once they had appeared, they flourished in immense numbers. An organism that produces a skeleton of stone and lives in an environment where deposits of ooze and sand are being laid down is an ideal subject for fossilisation. Huge thicknesses of limestone in many parts of the world consist almost entirely of coral remains and they provide a detailed chronicle of the development of the group.

The coral polyps secrete their skeletons from their bases. Each is connected with its neighbours by strands that extend laterally. As the colony develops, new polyps form, often on these connecting sections, and their skeletons grow over and stifle earlier polyps. So the limestone the colony builds is riddled with tiny cells where polyps once lived. The living ones form only a thin layer on the surface. Each species of coral has its own pattern of budding and so erects its own characteristic monument.

Corals are very demanding in their environmental requirements. Water that is muddy or fresh will kill them. Most will not grow at depths beyond the reach of sunlight for they are dependent upon single-celled algae that grow within their bodies. The algae photosynthesise food for themselves and in the process absorb carbon dioxide from the water. This assists the corals in the building of their skeletons, and releases oxygen which helps the corals respire.

The first time you dive on a coral reef is an experience never to be forgotten. The sensation of moving freely in three dimensions in the clear sunlit water that corals favour is, in itself, a bewitching and other-worldly one. But there is nothing on land that can prepare you for the profusion of shapes and colours of the corals themselves. There are domes, branches and fans, antlers delicately tipped with blue, clusters of thin pipes that are blood red. Some seem flower-like, yet when you touch them they have the incongruous scratch of stone. Often different coral species grow beside one another, mingled with sea pens arching above and beds of anemones that wave long tentacles in the current. Sometimes you swim over great meadows that consist entirely of one kind of coral; sometimes in deeper water, you discover a coral tower hung with fans and sponges that extends beyond your sight into depths of darkest blue.

But if you swim only during the day, you will hardly ever see the organisms that have created this astounding scene. At night, with a torch in your hand, you will find the coral transformed. The sharp outlines of the colonies are now misted with opalescence. Millions of tiny polyps have emerged from their limestone cells to stretch out their minuscule arms and grope for food.

Coral polyps are each only a few millimetres across, but, working together in colonies, they have produced the greatest animal constructions the world had seen long before humans appeared. The Great Barrier Reef, running parallel to the eastern coast of Australia for over 1,600 kilometres can be seen from the moon. So if, some 500 million years ago, astronauts from some other planet passed near the earth, they could easily have noticed in its

TWO

Building Bodies

The Great Barrier Reef swarms with life. The tides surging through the coral heads charge the water with oxygen and the tropical sun warms it and fills it with light. All the main kinds of sea animals seem to flourish here. Phosphorescent purple eyes peer out from beneath shells; black sea urchins swivel their spines as they slowly perambulate on needle tip; starfish of an intense blue spangle the sand; and patterned rosettes unfurl from holes in the smooth surface of coral. Dive down through the pellucid water and turn a boulder. A flat ribbon, striped yellow and scarlet, dances gracefully away and an emerald green brittle star careers over the sand to find a new hiding place.

The variety at first seems bewildering, but leaving aside primitive creatures like jellyfish and corals which we have already described, and the much more advanced backboned fish, nearly all can be allocated to one of three main types: shelled animals, like clams, cowries and sea snails; radially symmetrical creatures, like starfish and sea urchins; and elongated animals with segmented bodies varying from wriggling bristle worms to shrimps and lobsters.

The principles on which these three kinds of bodies are built are so fundamentally different that it is difficult to believe that they can be related to one another except right at the very roots of the evolutionary tree. The fossil record bears this out. All three groups, being sea-dwellers, have left behind abundant remains, and the details of their separate dynastic fortunes can be traced through the rocks for hundreds of millions of years. The walls of the Grand Canyon show that animals without backbones, invertebrates, came into existence long before the vertebrates such as fish. But just below the layer of gently folded limestones that contain the earliest of the invertebrate fossils, the strata change radically. Here the rocks are highly contorted. They had once formed mountains. These were eroded and eventually covered with the sea that deposited the limestone now lying above them. The episode occupied many millions of years and during all that time there were no deposits. As a consequence, this junction in the rocks represents a huge gap in the record. To trace the invertebrate lines back to their origins, we must find another site where rocks were not only deposited continuously throughout this critical period, but have survived in a relatively undistorted condition.

Such places are few, but one lies in the Atlas Mountains of Morocco. The bare hills behind Agadir in the west are built of blue limestones so hard that they ring under the fossil hunter's hammer. The beds of rock are slightly tilted but otherwise undistorted by earth movements. On the crest of the passes, the rocks yield fossils. They are not very many, but if you look hard enough you can collect quite a range of species. All fossils found anywhere in the world in rocks of this age can be placed in one or other of those three main groups we identified on the reef. There are tiny shells, the size of your little fingernail, called brachiopods; radially symmetrical organisms looking like stalked flowers called crinoids; and trilobites, segmented creatures that resemble woodlice.

The limestones at the top of the Moroccan succession are about 560 million years old. Beneath them lie more layers extending downwards for thousands of metres, seemingly unchanged in character. In them, surely, must be evidence about the origins of those three great invertebrate groups.

But it is not so. As you clamber down the mountain-side over the strata, the fossils suddenly disappear. The limestone seems to be exactly the same as that at the head of the pass, so the seas in which it was laid down must surely have been very similar to those that produced fossil-iferous rocks. There are no signs of a revolutionary change in physical conditions. It is simply that at one time the ooze covering the seafloor contained shells of animals – and before that it did not.

This abrupt beginning to the fossil record is not just a Moroccan phenomenon, though you can see it here more vividly than in most places. It occurs in almost all the rocks of this age throughout the world. The microfossils from the cherts of Lake Superior and South Africa show that life had started long, long before. In the theoretical year of life, shelled animals do not appear until early November. So the bulk of life's history is undocumented in the rocks. Only at this late date, about 600 million years ago, did several separate groups of organisms begin to leave records of any abundance by secreting shells. Why this sudden change should have come about, we do not know. Perhaps before this time the seas were not at the right temperature or did not have the chemical composition to favour the deposition of the calcium carbonate from which most marine shells and skeletons are constructed. Whatever the reason, we have to look elsewhere for evidence of the origins of the invertebrates.

We can find some living clues back on the reef. Fluttering over the coral heads, hiding in the crevices or clinging to the underside of rocks, are flat leaf-shaped worms. Like jelly-fish, they have only one opening to their gut through which they both take in food and eject waste. They have no gills and breathe directly through their skin. Their underside is covered with cilia which by beating enable them to glide slowly over surfaces. Their front end has a mouth below and a few light-sensitive spots above so that the animal can be said to have the beginnings of a head. These flatworms are the simplest creatures to show signs of such a thing.

Eye-spots, to be of any use, must be linked to muscles so that the animal can react to what it senses. In flatworms all that exists is a simple network of nerve fibres. There are a few thickenings in some of them, but these can hardly be described as brains. Yet the flatworms can learn the kind of things that would help even this simplest of animals to survive, such as avoiding a particularly dangerous place or remembering where food can be found.

Today we know of some 3,000 species of flatworm in the world. Most are tiny and water-living. You can find freshwater ones in most streams simply by dropping a piece of raw meat or liver into the water. If the under-water vegetation is thick, flatworms are likely to glide out in some numbers and settle on the bait. In humid tropical forests, the ground is usually moist enough for some species to live on land, and many are likely to appear, undulating on the mucus that they secrete from their undersides. One of these species grows to a length of about 60 centimetres. Other flatworms have taken to the parasitic life and live unseen within the bodies of other animals – including us.

Liver flukes still retain the typical flatworm form. Tapeworms are also members of the group, though they look very different, for after burying their heads in the walls of their host's gut, they bud off egg-bearing sections from their tail end. These segments remain attached while they mature, eventually forming a chain that may be as much as 10 metres long. The whole creature, as a result, looks as though it is divided into segments, but in fact these separate living packets of eggs are quite different from the permanent internal compartments of a truly segmented creature like an earthworm.

Flatworms are very simple creatures. Members of one free-swimming group lack a gut altogether and look very like the tiny free-swimming coral organisms before they settle down to a sedentary life. So there is little difficulty in believing those researchers who conclude from a study of the detailed structure of both adult and larva that the flatworms are descended from simpler organisms like corals and jellyfish.

During the period when these first marine invertebrates were evolving, between 600 and 1,000 million years ago, erosion of the continents was producing great expanses of mud and sand on the seabed around the continental margins. This environment must have contained abundant food in the form of organic detritus falling from the waters above as the single-celled organisms that floated in the surface waters died and drifted downwards. It also offered concealment and protection for any creature that lived within it. The flatworm shape, however, is not suited to burrowing. A tubular form is much more effective, and eventually worms with such a shape appeared. Some became active burrowers, tunnelling through the mud in search of food particles. Others lived half buried with their front end above the sediment. Cilia around their mouths created a current of water and from it they filtered their food.

Some of these creatures lived in a protective tube. In time, the shape of the top of this was modified into a collar with slits in it. This improved the flow of water over the tentacles. Further modification and mineralisation eventually produced a two-part protective shell around the front end. These were the first brachiopods, including *Lingulella*, an example of a species that has existed virtually unchanged for hundreds of millions of years.

The front end of a brachiopod is really quite complicated. Within the shell, it has a mouth surrounded by a group of tentacles. They are covered with beating cilia which create a current in the water. Any food particles in it are caught by the tentacles and then passed by them down to the mouth. While doing this, the tentacles perform another and important function, for the water brings with it dissolved oxygen which the animal needs in order to respire. The tentacles absorb it and so, in effect, they become gills. The shell enclosing the tentacles not only gives protection to these soft delicate structures, but concentrates the water into a steady stream so that it flows more effectively over them.

The brachiopods elaborated this design considerably over the next million years or so. One group developed a hole at the hinge end of one of the valves through which the worm-like stalk emerged to fasten the animal into the mud. This gave the shell the look of an upside-down Aladdin oil lamp, with the stalk as the wick, and so the group as a whole gained the name of lamp shell. The tentacles within the shell eventually became so enlarged that they had to be supported by delicate spirals of limestone.

There are other shelled worms to be found alongside the brachiopods in these ancient rocks. In one kind the elaborated worm did not attach itself to the seafloor but continued to crawl about and secreted a small conical tent of shell under which it could huddle when in danger. This was the ancestor of the most successful group of all these shelled worms, the molluscs, and it too has a living representative, a tiny organism called *Neopilina*, which was dredged up in 1952 from the depths of the Pacific. Today there are about 80,000 different species of molluscs with about as many again known from their fossils. You can find some of them in your garden; they are the snails and the slugs.

The lower part of the molluscan body is called the foot. Its owner moves itself about by protruding the foot from the shell and rippling its undersurface. Many species carry a small disc of shell on the side of it which, when the foot is retracted into the shell, forms a close-fitting lid to the entrance. The upper surface of the body is formed by a thin sheet that cloaks the internal organs and is appropriately called the mantle. In a cavity between the mantle and the central part of the body, most species have gills which are continually bathed by a current of oxygen-bearing water, sucked in at one end of the cavity and expelled at the other.

The shell is secreted by the upper surface of the mantle. One whole group of molluscs has single shells. The limpet, like *Neopilina*, produces shell at an equal rate right round the circumference of the mantle and so builds a simple pyramid. In other species, the front of the mantle secretes faster than the rear and creates a shell in a flat spiral, like a watch spring. In yet others, maximum production comes from one side so that the shell develops a twist and becomes a turret. The cowrie concentrates its secretion along the sides of the mantle, forming a shell like a loosely clenched fist. From the slit along the bottom, it protrudes not only its foot but two sections of its mantle which in life may extend over each flank of the shell and meet at the top. These lay down the marvellously patterned and polished surface characteristic of cowries.

The single-shelled molluscs feed not with tentacles within the shell like the brachiopods but with a radula, a ribbon-shaped tongue, covered with rasping teeth. Some use it to scrape algae from the rocks. Whelks have developed a radula on a stalk which they can extend beyond the shell and use to bore into the shells of other molluscs. Through the holes they have drilled, they poke the tip of the radula and suck out the flesh of their victim. Cone shells also have a stalked radula but have modified it into a kind of gun. They slyly extend it towards their prey – a worm or even a fish – and then discharge a tiny glassy harpoon from the end. While the tethered victim struggles, they inject a venom so virulent that it kills a fish instantly and can even be lethal to human beings. They then haul the prey back to the shell and slowly engulf it.

A heavy shell must be something of a handicap when actively hunting, and some carnivorous molluscs have taken to a faster if riskier life by doing without it altogether and reverting to the lifestyle of their flatworm-like ancestors. These are the sea slugs (nudibranchs) and they are among the most beautiful and highly coloured of all invertebrates in the sea. Their long soft bodies are covered on the upper side with waving tentacles of the most delicate colours, banded, striped and patterned in many shades. Though they lack a shell, they are not entirely defenceless, for some have acquired secondhand weapons. These species float near the surface of the water on their feathery extended tentacles and hunt jellyfish. As the sea slug slowly eats its way into its drifting helpless prey, the stinging cells of the victim are taken into its gut, complete and unsprung. Eventually these migrate within the sea slug's tissues and are concentrated in the tentacles on its back. There they give just the same protection to their new owners as they did to the jellyfish that developed them.

Other molluscs, such as mussels and clams, have shells divided into two valves lke those of a brachiopod and thus are known as bivalves. These creatures are much less mobile. The foot is reduced to a protrusion that they use to pull themselves down into the sand. For the most part, they are filter feeders, lying with valves agape, sucking water in through one end of the mantle cavity and squirting it out through a tubular siphon at the other. Since they do not need to move, great size is no disadvantage. Giant clams on the reef may grow to be a metre long. They lie embedded in the coral, their mantles fully exposed, a zigzag of brilliant green flesh spotted with black, which pulsates gently as water is pumped through it. They can certainly be quite big enough for a diver to put his foot into, but he would have to be very incautious indeed to get trapped. Powerful though the clam's muscles are, it cannot slam its valves shut. It can only heave them slowly together, and that gives plenty of notice of its intentions. What is more, even when the valves of a really large specimen are fully closed, they only meet at the spikes on the edge. The gaps between them are so big that if you plunge your arm through into the mantle, the clam is quite unable to grip it – though the experiment is a little less unnerving if it is tried first with a thick stick.

Some filter feeders like the scallops do manage to travel – by convulsively clapping their valves together and so making curving leaps through the water. By and large, however, adult bivalves live rather static lives and the spreading of the species into distant parts of the seabed is carried out by the young. The molluscan egg develops into a larva, a minuscule animated globule striped with a band of cilia, which is swept far and wide by ocean currents. Then, after several weeks, it changes its shape, grows a shell and settles down. The drifting phase of its life puts it at the mercy of all kinds of hungry animals, from other stationary filter feeders to fish, so in order that its species can survive, a mollusc must produce great numbers of eggs. And indeed it does. One individual may discharge as many as 400 million.

One branch of the molluscs, very early in the group's history, found a way of becoming highly mobile and yet retaining the protection of a large and heavy shell – they developed gas-filled flotation tanks. The first such creature appeared about 500 million years ago. Its flat-coiled shell was not completely filled with flesh as is that of a snail, but had its hind end walled off to form a gas chamber. As the animal grew, new chambers were added to provide sufficient buoyancy for the increasing weight. This creature was the nautilus, and we can get an accurate idea of how it and its family lived because a few nautilus species, just like *Lingulella* and *Neopilina*, have survived to the present day.

One of these species, the pearly nautilus, grows to about 20 centimetres across. A tube runs from the back of the body chamber into the flotation tanks at the rear so that the animal can flood them and adjust its buoyancy to float at whatever level it wishes. The nautilus feeds not only on carrion but on living creatures such as crabs. It moves by jet propulsion, squirting water through a siphon in a variation of the current-creating technique developed by its filter-feeding relatives. It searches for its prey with the help of small stalked eyes and tentacles that are sensitive to taste. Its molluscan foot has become divided into some ninety long grasping tentacles which it uses to grapple with its prey. In the centre of them it has a horny beak, shaped like that of a parrot and capable of delivering a lethal, shell-cracking bite.

About 400 million years ago, after some 100 million years of evolution, the nautiluses gave rise to a variant group with many more flotation chambers to each shell, the ammonites. These became much more successful than their nautilus relatives, and today their fossilised shells can be found lying so thickly that they form solid bands in the rocks. Those of some species grew as big as lorry wheels. When you find one of these giants embedded in the honey-coloured limestones of central England or the hard blue rocks of Dorset, you might think that such immense creatures could do little but lumber massively across the seabed. But where erosion has removed the outer shell, the elegant curving walls of the flotation chambers that are revealed remind you that these creatures may well have been virtually weightless in water and able, like the nautilus, to jet-propel themselves at some speed through the water.

About 100 million years ago, the ammonite dynasty began to dwindle. Perhaps there were ecological changes that affected their egg-laying habits. Maybe new predators had appeared. At any rate, many species died out. Other lines gave rise to forms in which the shells were loosely coiled or almost straight. One group took the same path as the sea slugs did in more recent times and lost their shells altogether. Eventually all the shelled forms except the pearly nautilus disappeared. But some shell-less ones survived and became the most sophisticated and intelligent of all the molluscs, the squids and cuttlefish and the octopus. These are the cephalopods.

The relics of the cuttlefish's ancestral shell can be found deep within it. This is the flat leaf of powdery chalk, the cuttlebone, that is often washed up on the seashore. The octopus has no trace of a shell even within the flesh of its body, but one species, the argonaut, secretes from one of its arms a marvellous paper-thin version shaped very like a nautilus shell but without chambers, which it uses not as a home for itself but as a delicate floating chalice in which to lay its eggs.

The squid and cuttlefish have many fewer tentacles than the nautilus – only ten – and the octopus, as its name makes obvious, has only eight. Of the three creatures, the squids are much the more mobile and have lateral fins running along their flanks which undulate and so propel the animal through the water. All cephalopods can, like the nautilus, use jet propulsion on occasion.

Cephalopod eyes are very elaborate. In some ways they are even better than our own, for a squid can distinguish polarised light, which we cannot do, and their retinas have a finer structure, which means, almost certainly, that they can distinguish finer detail than we can. To deal with the signals produced by these sense organs they have considerable brains and very quick reactions.

Some squids grow to an immense size. The aptly named colossal squid lives in the seas around Antarctica. It can reach nearly 100 kilos in weight and measure six metres from the end of its body to the tip of its outstretched tentacles. Its rival for the claim to be the largest species of all is the giant squid. The biggest so far discovered have in fact been slightly smaller and substantially lighter. Although there are records of even larger specimens of this species, it seems that these were not accurate. Nevertheless, we are unlikely to have discovered the biggest individuals of either species, so the record may yet be broken. The eyes of these huge cephalopods are even larger than might be expected. The biggest recorded were 27 centimetres across and are the largest known eyes of any kind of animal, five times bigger, for example, than those of the blue whale. Why the squid should have such gigantic eyes is a mystery.

It could be, however, that they need extremely sensitive eyes to detect the presence of their great enemy – the sperm whale. Squid beaks are often found in the stomachs of sperm whales, and their heads often carry circular scars with diameters that match a squid's suckers. So there seems little doubt that squids and whales regularly fight in the dark depths of the ocean. Maybe the squids' huge eyes help them to detect the presence of the only animal big enough to hunt them.

The intelligence of all the cephalopods – octopus, squid and cuttlefish – is well known. Octopus have been observed disguising themselves from an approaching enemy by covering themselves with shells or picking up two halves of a coconut and hiding within. Many species in all three groups have an extraordinary ability to change their colour and shape. They can camouflage themselves by matching almost any environment and also signal to one another with patterns and shapes that sweep across their bodies. A female squid has even been filmed signalling to a male lying alongside her that she is not ready to mate, while at the same time displaying a pattern on the other side of her body to summon another male. Octopus and squid, two of the most advanced animals in the ocean which least resemble human beings, are among the few, it seems, that can rival mammals in their intellectual abilities.

But what of the second great category of animals without backbones, the one represented in ancient rocks by the flower-like crinoids? As these are traced upwards through the rocks, they become more elaborate and their fundamental structure becomes clearer. Each has a central body, the calyx, rising from a stem like the seedhead of a poppy. From this sprout five arms which, in some species, branch repeatedly. The surface of the calyx is made up of closely fitting plates of calcium carbonate, as are the stems and branches. Lying in the rocks, the stems look like broken necklaces, their individual beads sometimes scattered, sometimes still in loose snaking columns, as though their thread had only just snapped. Occasionally gigantic specimens are found with stems 20 metres long.

they can suddenly swim away, writhing their five limbs like Catherine wheels.

The fivefold symmetry and the hydrostatically operated tube feet are such distinctive characteristics that they make other echinoderms very easy to recognise. The starfish and their more sprightly cousins, the brittle stars, both possess them. These creatures appear to be crinoids that have neither stalk nor rootlets and are lying in an inverted position with their mouths on the ground and their five arms outstretched. Sea urchins too are obviously related. They seem to have curled their arms up from the mouth as five ribs and then connected them by more plates to form a sphere.

The sausage-like sea cucumbers that sprawl on sandy patches in the reef are also echinoderms, although in most species their shelly internal skeletons are reduced to tiny structures beneath the skin. Most lie neither face up nor face down, but on their sides. At one end there is an opening called the anus, though the term is not completely appropriate for the animal uses it not only for excretion but also for breathing, sucking water gently in and out over tubules just inside the body. The mouth, placed at the other end, is surrounded by tube feet that have become enlarged into short tentacles. These fumble about in the sand or mud. Particles adhere to them and the sea cucumber slowly curls them back into its mouth and sucks them clean with its fleshy lips.

One highly specialised deep-sea sea cucumber, called a sea-pig, lives in the mud of the deep seabed at depths of up to 5,000 metres. They are rotund little creatures about

15 centimetres long and have large tube-like structures on their underside with which they rootle about in the mud. They have been filmed in the deep sea, assembled in herds, perhaps for reproduction or attracted by the smell of a new source of food drifting down from the surface.

If you pick up a sea cucumber, do so with care, for they have an extravagant way of defending themselves. They simply extrude their internal organs. A slow but unstoppable flood of sticky tubules pours out of the anus, fastening your fingers together in an adhesive tangle of threads. When an inquisitive fish or crab provokes them to such action, it finds itself struggling in a mesh of filaments while the sea cucumber slowly inches itself away on the tube feet that protrude from its underside. Over the next few weeks it will slowly grow itself a new set of entrails.

The echinoderms may seem, from a human point of view, to be a blind alley of no particular importance. Were we to imagine that life was purposive, that everything was part of a planned progression due to culminate in the appearance of the human species or some other creature that might rival us in dominating the world, then the echinoderms could be dismissed as of no consequence. But such trends are clearer in the minds of people than they are in the rocks. The echinoderms appeared early in the history of life. Their hydrostatic mechanisms proved a serviceable and effective basis for building a variety of bodies, but were not susceptible in the end to spectacular development. In the areas that suit them, they are still highly successful. A starfish on the reef can crawl across a clam, fasten its tube feet on either side of its gape and slowly wrench the valves apart to feed on the flesh within. The crown-of-thorns starfish occasionally proliferates to plague proportions and devastates great areas of coral. Crinoids are brought up in trawls from the deep sea several thousand at a time. If it is improbable that any further major developments will come from this stock, it is also unlikely, on the evidence of the last 600 million years, that the group will disappear as long as life remains possible at all in the seas of the world.

The third category of creatures on the reef contains those with segmented bodies. In this instance, we do have fossil evidence of even earlier forms than the trilobites found in the Moroccan hills. The Ediacaran deposits in Australia which contain the remains of jellyfish and sea pens also preserve impressions of segmented worms. One species, a 5-centimentre-long animal named *Spriggina* after Reg Spriggs who first discovered the Ediacara fossils, has a crescent-shaped head and up to forty segments, fringed on either side by leg-like projections. What exactly it was, nobody can agree. No legs have been identified, but this may be a limitation in the process of fossilisation. Some scientists think it may represent a completely extinct group. One widely accepted possibility is that it was a kind of annelid worm related to the bristle worms that are so common on a reef and the earthworms that you can find in your garden.

Annelids have grooves encircling their body that correspond to the internal walls that divide its interior into separate compartments. Each of these is equipped with its own set of organs. On the exterior and on either side, there are leg-like projections sometimes equipped with bristles, and another pair of feathery appendages through which oxygen is absorbed. Within its body, each segment has a pair of tubes opening to the exterior from which waste is secreted. A gut, a large blood vessel and a nerve cord run from front to end through all the segments, linking and coordinating them.

Fossils can only tell us so much. Even the exceptionally well-preserved remains of Ediacara offer no clue about the connection between the segmented worms and the other early groups. However, there is one further category of evidence to be looked at – the larvae. The segmented worms have spherical larvae with a belt of cilia round their middles and a long tuft on top. These are almost identical to the larva of some molluscs, a strong indication that back in time the two groups sprang from common stock. The echinoderms, on the other hand, have a larva that is quite different, with a twist to its structure and winding bands of cilia around it. This group must have separated from the ancestral flatworms at a very early stage indeed, long before the split between the molluscs and the segmented worms. Geneticists, analysing the DNA of each of these groups, now confirm these deductions and reveal that there are two major groupings of bilaterally symmetrical animals. Octopus, crabs and flatworms form one group, while echinoderms, tunicates and all the backboned animals make up the other.

Segmentation may have developed as a way of enabling worms to increase their efficiency as burrowers in mud. A line of separate limbs down each side is clearly a very effective structure for this purpose and it could have been acquired by repeating the simple body unit to form a chain. The change must have taken place long before Ediacaran times, for when those rocks were deposited the fundamental invertebrate divisions were already established The Ediacaran fossils, in Australia where they were first discovered and in Britain, Newfoundland, Namibia and Siberia, now confirm these deductions. Thereafter their history remains virtually invisible for a 100 million years. Only after this vast span do we reach the period, 600 million years ago, represented by the Moroccan deposits and others throughout the world. By that time many organisms had, as we have seen, developed shells from which we can deduce their existence and shape, but not much more.

However, there is one exceptional fossil site dating from only a little later than those of Ediacara that provides far more detailed information about the bodies of animals than can come from mere shells. In the Rocky Mountains of British Columbia, the Burgess Pass crosses a ridge between two high snowy peaks. Close to its crest lies an outcrop of particularly fine-grained shales, and in these have been discovered some of the most perfectly preserved fossils in the world. The shales were laid down about 530 million years ago, close to the beginning of the Cambrian period in a basin of the seafloor at a depth of about 150 metres. It must have been sheltered by a submarine ridge, for there were no currents to disturb the fine sediments on also small disc-shaped fossils marked with lines radiating from its centre that looked somewhat like a tiny slice of pineapple, which was initially thought to be some kind of jellyfish. Perhaps strangest of all, there was an elongated segmented animal that appeared to have seven pairs of spiny stilt-like legs, and seven flexible tentacles along its back, each ending in a tiny mouth. It seemed so strange as to be almost nightmarish, and the researcher who studied it accordingly called it *Hallucigenia*.

Subsequent work, however, showed that these oddities were not the founder members of some wholly unsuspected animal groups. A very exceptional specimen of *Anomalocaris* showed that the 'strange shrimps' were not complete animals but just the forelimbs belonging to a much bigger creature that used them to grab its prey. And the pineapple slice was eventually shown to have in its centre minute teeth. It was a mouth that belonged to the same animal as the tentacles. Both these pieces of *Anomalorcaris'* body apparently had a more heavily strengthened exoskeleton and so regularly became separated from the animal's more easily decayed body. As for *Hallucigenia*, further research on other specimens showed that it had been reconstructed in an upside-down position. The spindly legs were in fact protective dorsal spines, and what had been considered tentacles were in reality its legs. It is now thought that it may be the first known member of a strange group called the lobopods which today includes odd little creatures called velvet worms.

The great variety of creatures in the Burgess Shales is a reminder of how incomplete our knowledge of all fossil faunas actually is. The ancient seas contained many more kinds of animals than we can ever know. In this one site, conditions allowed a uniquely large proportion to be preserved, but even this is only a hint of what must have once existed.

The Burgess Shales also contain superbly preserved examples of trilobites like those in the Moroccan limestones. Their body armour was constructed partly of calcium carbonate and strengthened by a horny substance called chitin, a material that forms the external skeletons of insects. But chitin, unlike skin, does not expand, so any animal with such an external chitinous skeleton has to shed it regularly if it is to grow – as indeed insects do today. Many of the trilobite fossils we find are in fact these empty suits of armour. Sometimes they are concentrated in great drifts, having been sorted by sea currents, as shells sometimes are when they are swept up on beaches today. The underwater avalanches in the Burgess Shales Basin, however, swept down not just discarded armour but living trilobites and buried them. Mud particles filtered into the animals' bodies and preserved the finest details of their anatomy. So in them we can still see the paired jointed legs that are attached to each body segment, the feathery gill associated with each leg, two feelers at the front of the head, and the gut running the entire length of the body. Even the muscle fibres along the back, which enabled the animal to roll itself up into a ball, are still recognisable in some exceptional specimens.

Trilobites, as far as we know, were the first creatures on earth to develop high-definition eyes. They are mosaics, a cluster of separate components, each with its own lens of crystalline calcite orientated in the precise position in which it transmits light most efficiently, much like the eyes of today's insects. One eye may contain 15,000 elements, and would have given its owner an almost hemispherical field of view. Late in the dynasty, some species developed an even more sophisticated kind of eye and one that has never been paralleled by any other animal. Here the components are fewer but larger. Their lenses are much thicker and it is thought that these species lived where there was little light and needed thick lenses to collect and concentrate what light there was. However, the optical properties of a simple calcite lens in contact with water are such that it transmits light in a diffused way and cannot bring it to a sharply focused point. To do this, a two-part lens is needed which has a waved surface at the junction between its two elements. And this is exactly what these trilobites evolved. The lower element of the double lens was formed by chitin and the surface between the two conforms to the mathematical principle that human scientists did not discover until 300 years ago when they tried to correct the spherical aberration of lenses in their newly invented telescopes.

As the trilobites spread through the seas of the world, they diversified into a great number of species. Many seem to have lived on the seafloor, chomping their way through mud. Some colonised the deep seas where there was little light and lost their eyes altogether. Others, to judge from the shape of their limbs, may well have paddled about, legs uppermost, scanning the seafloor below with their large eyes.

In due course, as creatures of many kinds and varying ancestries came to live on the bottom of the seas, the trilobites lost their supremacy. Two hundred and fifty million years ago, their dynasty came to an end. One relation alone survives, the horseshoe crab. It's a misleading name for it is not a crab and only half its shell bears any resemblance to a horseshoe. Measuring 30 centimetres or so across, it is many times bigger than most known trilobites and its armour no longer shows any signs of segmentation. Its front section is a huge domed shield, on the front of which are two bean-shaped compound eyes. A roughly rectangular plate, hinged to the back of the shield, carries a sharp spike of a tail. But beneath its shell, the animal's segmentation is clear. It has several pairs of jointed legs with pincers on the end, and behind these there are plate-like gills, large and flat like the leaves of a book.

Horseshoe crabs are seldom seen, for they live at considerable depths. Some inhabit Southeast Asian waters, others are found in the seas along the North Atlantic coast of America. Every spring, they migrate towards the coast. Then on three successive nights, when the moon is full and the tides are high, hundreds of thousands emerge from the sea. The females, their huge shells glinting in the moonlight, move towards the shore, dragging smaller males behind them. Sometimes four or five males, in their anxiety to reach a female, cling to one another and form a chain. As she reaches the edge of the water, the female half buries herself in the sand. There she sheds her eggs and the males release sperm. For kilometre after kilometre along the dark beaches, the living tide of horseshoe crabs is so thick that they form a continuous strip, like a causeway of giant cobbles. The breakers sometimes overturn them and they lie in the sand, with their legs waving, their stiff tails slowly swivelling, in an effort to lever themselves right side up. Many fail and are abandoned by the receding tide to die as thousands more swim in the shallows, pressing forward to take their turn.

This scene must have been enacted every spring for several hundred million years. When it began, the land was without life of any kind, and on such beaches the eggs were safe from sea-dwelling marauders. Perhaps this is why the horseshoe crabs developed the habit. Today beaches are not quite so safe, for hordes of gulls and small wading birds congregate to share the prodigious feast. But many of the fertilised eggs remain buried deep among the sand grains where they will stay for a month until, once more, high water reaches this part of the beach, stirring the sand and releasing the larvae to swim freely in the sea.

Although the trilobites were so successful, they were by no means the only armoured creatures to develop from the segmented worms. So did a group that must have been among the most alarming of all marine monsters – the sea scorpions, called scientifically the Eurypterids. Some grew to a length of two metres and were the largest arthropods ever known to have existed. However, in spite of their appearance and huge claws, many of them were filter feeders. Presumably, their fearsome claws were used in fights between one another rather than in subduing prey. Like the trilobites, they disappeared at the end of the Permian period.

One group related to the trilobites did however survive and today is extremely successful. They differed in one seemingly trivial but nonetheless diagnostic characteristic. They have not one but two pairs of antennae on their heads. They lived alongside the trilobites, comparatively unobtrusively for hundreds of millions of years, and then, when the trilobite dynasty came to an end, it was they who took over. They are the crustaceans. Today there are about 35,000 species of crustacean – seven times as many as there are of birds. Most prowl among the rocks and reefs – crabs, shrimps, prawns and lobsters. Some – the barnacles – have taken up a static life. Others – the krill which forms the food of whales – swim in vast shoals.

An external skeleton is highly versatile; it serves the tiny water flea as well as it does the giant Japanese spider crab that measures over three metres from claw to claw. Each crustacean species modifies the shape of its many paired legs for particular purposes. Those at the front may become pincers or claws; those in the middle, paddles, walking legs or tweezers. Some have feathery branches, gills through which oxygen is absorbed from the water. Others develop attachments so that they can carry eggs.

The limbs, which are tubular and jointed, are operated by internal muscles. These extend from the end of one section, along its length, to a prong from the next section which projects across the joint. When the muscle contracts between these two attachment points, the limb hinges. Such joints can only move in one plane, but crustaceans deal with that limitation by grouping two or three on a limb, sometimes close together, each working in a different plane so that the free end of the limb can move in a complete circle.

The external shell, however, gives the crustaceans the same problem as it gave the trilobites. It will not expand, and since it completely encloses their bodies, the only way they can grow is to shed it periodically. As the time for the moult approaches, the animal absorbs much of the calcium carbonate from its shell into its blood. It secretes a new, soft and wrinkled skin beneath the shell. The outgrown armour splits at the back and the animal pulls itself out, leaving the shell more or less complete, like a translucent ghost of its former self. Now, because the animal's skin is soft, it must hide, but it grows fast and swells its body by absorbing water and stretching out the wrinkles of its new carapace. Gradually this hardens so the animal can again venture into a hostile world.

The hermit crab partly avoids this complicated and hazardous process by having a shell-less hind part and protecting it with a discarded mollusc shell, slipping into a new one in a minute or so whenever it has the need.

The external skeleton has one incidental quality which has had momentous results. Mechanically, it works almost as well on land as it does in water, so that, providing a creature can find a way of breathing, there is little to prevent it walking straight out of the sea and up the beach. Many crustaceans, indeed, have done so – sand shrimps and beach hoppers stay quite close to the sea; and pill bugs and penny sows have colonised moist ground throughout the land.

The most spectacular of all these land-living crustaceans is the robber crab. It is found on islands in the Indian

Ocean and the western parts of the Pacific. At the back of its main carapace, at the junction with the first segment of its abdomen, there is an opening to an air chamber lined with moist puckered skin through which the animal absorbs oxygen. This monster is so big it can embrace the trunk of a palm tree between its outstretched legs. It climbs with ease, and once in the palm's crest, cuts down with its gigantic pincers the young coconuts on which it feeds. It has to return to the sea to lay its eggs, but otherwise it is entirely at home on land.

Other descendants of the marine invertebrates have also left the water. Among the molluscs there are the snails and the shell-less slugs, but these emerged from water relatively recently in the group's history. The first to make the move to land were probably descendants of the segmented worms, the millipedes. Their droppings have been found fossilised in the rocks of Shropshire. They were followed by pioneers which recent DNA studies show to have been crustaceans. And some of these made such a success of life in their new surroundings that they eventually gave rise to the most numerous and diverse group of all land animals – the insects.

THREE

The First Forests

There are few more barren places on earth than the plains surrounding a volcano in the aftermath of its eruption. Black tides of lava lie spilt over its flanks like slag from a furnace. Their momentum has gone but they still creak, and boulders still tumble as the flow settles. Steam hisses between the blocks of lava, caking the mouths of the vents with yellow sulphur. Pools of liquid mud, grey, yellow or blue, boiled by the subsiding heat from far below, bubble creamily. Otherwise all is still. No bush grows to give shelter from the scouring wind; no speck of green relieves the black surface of the empty ash plains.

This desolate landscape has been that of much of the earth for the greater part of its history. The first volcanoes to appear on the surface of the cooling planet erupted on a far greater scale than any that we know today, building entire mountain ranges of lava and ash. Over the millennia, the wind and rain destroyed them. Their rocks weathered and turned to clay and mud. Streams transported the debris, particle by particle, and strewed it over the seafloor beyond the margins of the land. As the deposits accumulated, they compacted into shales and sandstone.

The continents were not stationary. They drifted slowly over the earth's surface, driven by the convection currents moving deep in the earth's mantle. When they collided, the sedimentary deposits around them were squeezed and rucked up to form new mountain ranges. As the geological cycles repeated themselves for some three thousand million years, and the volcanoes exploded and spent themselves, the land remained barren. In the sea, however, life burgeoned.

Some marine algae no doubt managed to live on the edges of the seas, rimming the beaches and boulders with green, but they could not have spread far beyond the splash zone, for they would have dried out and died. Then between 450 and 500 million years ago, some forms developed a waxy covering, a cuticle, which warded off desiccation. Even this, however, did not totally emancipate them from water. They could not leave it because their reproductive processes depended on it.

Algae reproduce themselves in two ways – by straightforward asexual division and by the sexual method, which is of great importance in the the evolutionary process. Sex cells will only develop further if they meet each other and fuse in pairs. To make these journeys and achieve these meetings, they need water.

This problem still besets the most primitive land plants living today – both the flat, moist-skinned ones known as liverworts, and the filamentous ones covered with green scales, the mosses. They use these two methods of reproduction, sexual and asexual, in alternate generations. The familiar green moss is the generation which produces the sex cells. Each large egg remains attached to the top of the stem, while the smaller microscopic sperms are released into water and wriggle their way up to fertilise it. The egg then germinates while still attached to the parent plant and produces the next asexual generation – a thin stem with, at its tip, a hollow capsule. In this, great numbers of grain-like spores are produced. When the atmosphere becomes dry, the capsule wall expands until it suddenly snaps apart, throwing the spores into the air to be distributed by the wind. Those that land on a suitably moist site then develop into new plants.

Moss filaments have no rigidity. Some kinds achieve a modest height by packing closely together in cushions and so giving one another support, but their soft, permeable, water-filled cells do not provide enough strength to enable individual stems to stand upright. Plants like these are very likely to have been among the earliest forms to colonise the moist margins of the land, but so far no fossil relics of undoubted mosses have been discovered from this early period.

The first land plants we have indentified, dating from over 400 million years ago, are simple leafless branching strands which occur as filaments of carbon in the rocks of central Wales and in some cherts in Scotland. Like mosses, they had no roots, but when their stems are carefully prepared and examined under the microscope, they are seen to contain structures that no moss possesses – long, thick-walled cells that must have conducted water up the stem. These structures gave them strength and enabled them to stand several centimetres tall. That may not sound very imposing, but it represented a major advance in life's colonisation of the land.

Such plants, together with primitive mosses and liverworts, formed green tangled carpets, miniature forests that spread inland from the edges of estuaries and rivers, and into these crept the first animal colonists from the sea. They were segmented creatures, ancestors of today's millipedes, well suited by their chitinous armour to movement on land. At first they doubtless kept close to the edge of the water, but wherever there was moss there was both moisture and vegetable debris and spores to eat. With the land to themselves, these pioneering creatures flourished. Their name millipede, 'thousand legs', is something of an overstatement. No species alive today has many more than two hundred legs, and some have as few as eight. Nevertheless, some of them grew to magnificent dimensions. One of them was two metres long and must have had a devastating effect on the plants as it browsed its way through the wet green bogs. It was, after all, as long as a cow.

The external skeleton inherited from their water-living forebears needed few modifications for life on land, but the millipedes did have to acquire a different method of breathing. The feathery gill attached to a stalk alongside the leg that had served their aquatic relatives, the crustaceans, would not work in air. In its place, the millipedes developed a system of breathing tubes, the tracheae. Each tube begins at an opening on the flank of the shell and then branches internally into a fine network that leads ultimately to all the organs and tissues of the body, the tips even entering individual specialised cells called tracheoles that deliver gaseous oxygen to the surrounding tissues and also absorb waste.

Reproduction out of water, however, posed problems for the millipedes. Their marine ancestors had relied, like the algae, on water to enable their sperm to reach their eggs. On land the solution was an obvious one – male and female, being well able to move about, must meet and transfer the sperm directly from one to the other. This is exactly what millipedes do. Both sexes house their reproductive cells in glands close to the base of the second pair of legs. When the male meets the female in the mating season, the two intertwine. The male reaches forward with his seventh leg, collects a bundle of sperm from his sex gland and then clambers alongside the female until the bundle is beside her sexual pouch and she is able to take it in. The process looks rather laborious but at least it is not dangerous. Millipedes are entirely vegetarian. Fiercer invertebrates, which came to the moss jungles to prey on this grazing population of millipedes, could not indulge in such trusting relationships.

Three groups of these predatory creatures still survive today – centipedes, scorpions and spiders. Like their prey, they are members of the segmented group of animals, though the degree to which they have retained divisions in their bodies varies considerably. The centipedes are as clearly and extensively segmented as their close relatives the millipedes. The scorpions show divisions only in their long tails; and most spiders have completely lost all signs of segmentation, except for a few Southeast Asian species which retain clearly recognisable relics of their segmented past.

The scorpions that live today have not only fearsome-looking claws but a large venom gland drooping from the end of a long thin tail with a sharp curving sting. Their copulations cannot be the somewhat hit-and-miss gropings practised by the millipedes. Approaching such an aggressive and powerful creature is a dangerous enterprise even if the move is made by another individual of the same species and its intentions are purely sexual. There is a real risk of it being regarded not as a mate but a meal. So scorpion mating demands, for the first time among the animals that have appeared so far in this history, the ritualised safeguards and placations of courtship.

The male scorpion approaches the female with great wariness. Suddenly he grabs her pincers with his. Thus linked, with her weapons neutralised, the pair begin to dance. Backwards and forwards they move with their tails held upright, sometimes even intertwined. After some time, their shuffling steps have cleared the dancing ground of much of its debris. The male then extrudes a packet of sperm from the genital opening beneath his thorax and deposits it on the ground. Still grasping the female by the claws, he jerks and heaves her forward until her sexual opening, also on her underside, is brought directly above the sperm packet. She takes it up, the partners disengage and then go their separate ways. The eggs eventually hatch inside the mother's pouch, the young crawl out and clamber up on to her back. There they stay for about a fortnight until they have completed their first moult and can fend for themselves.

Spiders, too, must be extremely cautious in their court-ship. Matters are made even more hazardous for the male because he is nearly always smaller than the female. And he prepares for his encounter with his mate long before he meets her. He spins a tiny triangle of silk a few millimetres in length and deposits a drop of sperm on to it from the gland that lies underneath his body. He then sucks it into the hollow first joint of his pedipalp, a special limb at the front of his body. Now he is ready.

The courtships of spiders are beguilingly various and ingenious. Jumping spiders and wolf spiders hunt primarily by sight and have excellent eyes. The courting male, conse-quently, relies on visual signals to make the female aware of his presence and his purpose. His pedipalps are brightly coloured and patterned, and as soon as he sights a female, he begins to signal with them in a kind of manic semaphore. Nocturnal spiders, on the other hand, depend largely on an extremely delicate sense of touch to find their prey. When they meet one another, they gingerly caress each other's long legs, and only after a great deal of hesitation do they come to closer quarters. Web-making spiders are sensitive to the vibrations on their silken threads that tell them when a victim has blundered into the web. So when the male of such a species approaches a female hanging, large and menacing, on her web, or lurking hidden beside it, he signals to her by twanging the threads at one side in a special and meaningful way which he trusts the female will recognise. Other species put their faith in bribery. The male catches an insect and carefully parcels it up in silk. Holding this in front of him, he cautiously approaches the female and presents it to her. While she is occupied in examining the gift, he quickly scuttles over her and ties her to the ground with bonds of silk. Only then does he risk an embrace.

All these techniques lead to the same conclusion. The male, having survived every danger, places his pedipalp close to the female's genital opening, squirts out the sperm and then hastily retreats. It has to be recorded that in spite of all his precautions he sometimes fails to make his getaway in time and the female eats him after all. But in terms of the transmission of his genes, the male's disaster is of limited consequence: he lost his life after, not before, he had completed his purpose.

While the early segmented animals were perfecting their adaptations for living on land and away from moisture, the plants were also changing. Neither the mosses nor the other early forms had true roots. Their short upright stems sprang from a horizontal one of a similar character lying along the ground or just below it. This construction served well enough in moist surroundings, but in many parts of the world the only permanent water supply lies below the ground. To tap that requires roots that probe deep between the particles of the soil and can absorb the film of water that clings to them in all except the most arid environments. Three groups of plants appeared that possessed such structures, and all three have descendants that have survived without much change: club mosses, which resemble mosses but have stiffer stems; horsetails, which grow in waste patches and ditches and have stems encircled at intervals with rings of needle-like leaves; and ferns.

The ferns, early in their history, had developed a special protein to protect themselves from damage by ultraviolet light, something that had not been a problem for their ancestors since they lived in water where such wavelengths could not reach them. This substance now slowly changed into a material called lignin. This is the basis of wood, and it gave them the rigidity needed to grow tall. So a new kind of competition developed between plants.

All green plants depend on light to power the chemical processes by which they use simple elements to synthesise their body substances. So if a plant does not grow tall, it risks being overshadowed by its neighbours and condemned to shade where, starved of light, it might die. So these early groups used the newly acquired strength of their stems to grow very tall indeed. They became trees. The club mosses and horsetails were still, for the most part, swamp-dwellers, and there they now stood in dense ranks, thirty metres tall, some with woody trunks two metres in diameter. The compacted remains of their stems and leaves today form coal. The great thicknesses of the seams are impressive evidence of the abundance and persistence of the early forests. Other species of both these groups also spread farther inland and there mingled with ferns. These had developed true leaves, large spreading structures with which to collect as much light as possible. They grew tall with curving trunks, like the tree ferns that still thrive in tropical rainforests.

The height of these first forests must have caused considerable problems for their animal inhabitants. Once, there had been a superabundance of leaves and spores close to the ground. Now the soaring trunks had raised this source of food high in the sky, creating a dense canopy that cut out much of the light. The floor of these forests was, at best, only sparsely vegetated and great areas may have been entirely without any living leaves. Some of the multi-legged vegetarians found their fodder by clambering up the trunks.

There may have been another factor that induced these creatures to leave the ground. About this time, animals of a completely new kind joined the invertebrates on the land. They had backbones and four legs and wet skins. They were the first amphibians and they too were carnivorous. A description of their origins and fate will have to wait until we have followed the development of the invertebrates to its climax, but their presence at this stage must be mentioned if the scene in these first jungles is not to be misrepresented.

Virtually all of the new-style invertebrate families still survive. Among the most numerous are the bristletails and springtails. Although they are little known and infrequently seen, they are enormously abundant. There is hardly a spadeful of soil or leaf litter anywhere in the world that does not contain some of them. Indeed, the springtails, or collembola, are probably the most abundant arthropods on the planet. Most are only a few millimetres long. Of those new families, only one is commonly noticed – the silverfish that glides smoothly across cellar floors or is occasionally discovered making a meal of the dried glue in the bindings of books. Its body is clearly segmented but it has very many fewer divisions than the millipede. It has a well-defined head with compound eyes and antennae; a thorax bearing three pairs of legs, the result of fusing together three segments; and a segmented abdomen which, while it no longer carries limbs on each segment, retains little stumps as signs that it once possessed them. Three thin filaments trail from its rear end. It breathes like the millipedes by means of tracheae, and it reproduces in a manner reminiscent of those early land invertebrates, the scorpions. The male silverfish deposits a bundle of sperm on the ground and then, one way or another, he entices the female to walk over it. When that happens, she is stimulated to take it up into her own sexual pouch.

There are several thousand different species of bristletails and collembola. They vary considerably in their anatomy and, as is often the case when considering the simpler members of a big group, it is sometimes difficult to decide whether a particular characteristic represents a truly primitive survival or one that has become secondarily reduced to suit a particular way of life. The silverfish, for example, has compound eyes but other members of the group are blind. All lack wings. Some even lack tracheae and breathe through their chitinous skeleton which is particularly thin and permeable. Is this because they never had them or because they have lost them?

Many such debatable questions raised by the anatomy of these creatures still wait universally agreed answers. However, they all have six legs and tripartite bodies and these characteristics clearly link them to that great and varied group of land invertebrates, the insects. They appeared many millions of years after the earlier groups were well established. Geneticists have now shown that collembolla,as well as the insects, including the silverfish, are all closely related to one particular group of water-living crustaceans, the remipedia (the name means 'oar-foot'), which today are found only in the pools and streams of caves.

The primitive insects must have found some of their food by climbing the trunks of the early tree ferns and horsetails. The ascent was doubtless relatively easy. The climb down, involving long detours over the upward-pointing leaf-bases, may have been very much more laborious and time-consuming. Whether or not the prevalence of such obstacles had anything to do with the next developments, we cannot be sure. It is certain, however, that some of these primitive insects did develop a much swifter and less laborious method of getting down. They flew.

We have no direct evidence of how they achieved flight, but the living silverfish provides a clue. On the back of its thorax it has two flap-like sideways extensions of the chitinous shell that look as though they might be the rudiments of wings. The early wings may not have served initially for flight. Insects, like all animals, are greatly affected by body temperature. The warmer they are, the quicker the energy-producing chemical reactions of their body can proceed and the more active they can be. If their blood were to be circulated through thin flaps extending laterally from the back, they could certainly warm themselves very effectively and quickly in the sunshine. If, furthermore, these flaps had muscles at their base, they could be tilted to face squarely to the sun's rays. Insect wings do originate as flaps on the back and they do, initially, have blood flowing in their veins, so such a theory seems very plausible.

However this may be, insects with wings appeared some 350 million years ago. The earliest so far discovered are dragonflies. There were several species, most about the size of those living today. But for the dragonflies as for millipedes and other groups that have pioneered a new environment, the absence of competition allowed some early forms to develop to an enormous size, and dragonflies eventually appeared with a wingspan of 70 centimetres, the largest insects ever to exist. When the air became more thickly populated, such extravagant forms disappeared.

Living dragonflies have two pairs of wings which have simple joints to them: they can only move up and down and cannot be folded back. Even so, they are highly accomplished flyers, shooting over the surface of a pond in a blur of gauzy wings at up to 30 kph. At such speeds, they need accurate sense organs if they are to avoid damaging collisions. A tuft of hair on the front of the body helps them to check that their motion through the air is straight, but their primary navigational guidance comes from huge mosaic eyes on either side of the head, which provide superbly accurate and detailed vision.

Because of this dependence on sight, most dragonflies do not fly at night, although there are some that migrate vast distances over the oceans, flying from India to Africa and stopping off at the islands of the Maldives along the way. All are daytime hunters, flying with their six legs crooked in front of them to form a tiny basket in which they catch smaller insects. That fact alone makes it clear that they must have been preceded into the air by other herbivorous forms which, judging from the primitive nature of their anatomy, were probably related to cockroaches, grasshoppers, locusts and crickets.

The presence of these large populations of insects, whirring and buzzing through the air of the ancient forests, was eventually to play an extremely important part in a revolution that was taking place among the plants.

The early trees, like their predecessors, the mosses and liverworts, existed in two alternating forms, a sexual generation and an asexual one. Their greater height posed no problem for spore dispersal: if anything it was a help, since up in the treetops spores were more easily caught by the wind and carried away. The distribution of sex cells, however, was a different matter. Hitherto, it had been achieved by the male cells swimming through water, a process which demanded that the sexual generation be small and live close to the ground. That of ferns, club mosses and horsetails still is. The spores of these plants develop into a thin filmy plant called the thallus which looks not unlike a liverwort and releases its sex cells from its underside where there is permanent moisture. After its eggs have been fertilised, they grow into tall plants like the previous spore-producing generation.

On the ground, the thallus is clearly vulnerable. It is easily grazed by animals; if it dries out it dies; and the very success of the asexual generation with their arching fronds cuts it off from life-giving light. Many advantages would follow if it too could grow tall, but this would require a new technique for getting the male cell to the female.

There were two mechanisms available – the ancient, rather hazardous and capricious method that distributed spores, the wind; and the newly arrived messenger service, the flying insects, which were now regularly moving from tree to tree, feeding on the leaves and the spores. Plants took advantage of both mechanisms. About 350 million years ago, some appeared in which the sexual generation no longer grew flat on the ground, but up in the crowns of the trees. One group among these plants, the cycads, survives today and shows the development at a particularly dramatic stage.

Cycads look superficially like ferns, with long coarse feathery fronds. Some individuals produce tiny spores of the ancient type that can be distributed by the wind. Others develop much larger ones. These are not blown away but remain attached to the parent. There they develop the equivalent of the thallus, a special kind of conical structure within which eggs eventually appear. When a wind-blown spore – which now can be called pollen – lands on an egg-bearing cone, it germinates, not into a filmy thallus for which there is now no need, but into a long tube which burrows its way down into the female cone. The process takes several months. Eventually, when the tube is complete, a sperm cell is produced from the end of the tube. It is a majestic ciliated sphere, the largest known sperm of any organism, plant or animal, so big that a single one is visible to the naked eye. Slowly it makes its way down the tube. When it reaches the bottom, it enters a small drop of water that has been secreted by the surrounding tissues of the cone. There it swims, slowly spinning, driven by its cilia, as it re-enacts in miniature the journeys made through the primordial seas by the sperm cells of its algal ancestors. Only after several days does it fuse with the egg and so complete the long process of fertilisation.

Another group of plants adopting a similar strategy to the cycads arose at about the same time. These were the conifers – pines, larches, cedars, firs and their relations. They too rely on the wind to distribute their pollen. Unlike the cycads, they produce both pollen and egg-bearing cones on the same tree. The process of fertilisation in a pine takes even longer. The pollen tube requires a whole year to grow down and reach the egg, but once there, it contacts the egg cell directly, and the male cell, after descending the tube, does not tarry in a drop of water but fuses directly with the egg. The conifers have at last eliminated water as a transport medium for their sexual processes.

They have also developed one further refinement. The fertilised egg remains in the cone for one more year. Rich food supplies are laid down within its cells and waterproof coats are wrapped round it. Eventually, more than two years after fertilisation began, the cone dries and becomes woody. Its segments open, and out drop the fertilised fully provisioned eggs – seeds – which if necessary can wait for years before moisture penetrates them and stimulates them to spring to life.

By any standards, the conifers are a great success. Today, they constitute about a third of the forests of the world. The biggest living organism of any kind is a conifer, the giant redwood of California, which grows to 100 metres in height. Another conifer, the bristle-cone pine, which grows in the dry mountains of the southwestern United States, has one of the longest life-spans of any individual organism. The age of trees can easily be calculated if they grow in an environment where there are distinct seasons. In summer, when there is plenty of sunshine and moisture, they grow quickly and produce large wood cells; in winter, when growth is slow, the cells are smaller and the wood consequently more dense. This produces annual rings in the trunk. Counting those in the bristle-cone pine establishes that some of these gnarled and twisted trees germinated over five thousand years ago at a time when people in the Middle East were just beginning to invent writing, and the trees have remained alive throughout the entire duration of civilisation.

Conifers protect their trunks from mechanical damage and insect attack with a special gummy substance, resin.

When it first flows from a wound it is runny, but the more liquid part of it, turpentine, quickly evaporates, leaving a sticky lump which seals the wound very effectively. It also, incidentally, acts as a trap. Any insect touching it becomes inextricably stuck and very often buried as more resin flows around it. Such lumps have proved to be the most perfect fossilising medium of all. They survive as pieces of amber and contain ancient insects in their translucent golden depths. When the amber is carefully sectioned, it is possible, through the microscope, to see mouthparts, scales and hairs with as much clarity as if the insect had become entangled in the resin only the day before. Scientists have even been able to distinguish tiny parasitic insects, mites, clinging to the legs of the bigger ones. Extracting the DNA from a blood-sucking arthropod seems likely to remain science fiction, however. Even attempts to do so from insects trapped in copal, the modern equivalent of amber only a few decades old, have all met with failure.

The oldest pieces of amber so far discovered date from around 230 million years ago, a long time after the conifers and the flying insects first appeared, but they contain a huge range of creatures, including representatives of all the major insect groups that we know today. Even in the earliest specimens, each type has already developed its own characteristic way of exploiting that major insect invention, flight.

The dragonflies beat their two wings synchro-nously, with the front pair raised while the rear pair are lowered. This, however, creates very considerable physiological complexities. Their wings do not normally come into contact, but even so there are problems when the dragonfly executes sharp turns. Then the fore- and hindwings, bending under the additional stress of the turn, beat against one another, making an audible rattle that you can easily hear as you sit watching them make their circuits over a pond.

The later insect groups seem to have found that flight was more efficiently achieved with just one pair of beating membranes. Bees and wasps, flying ants and sawflies all hitch their fore- and hindwings together with hooks to make, in effect, a single surface. Butterfly wings overlap. Hawkmoths, which are among the swiftest insect flyers, capable of speeds of 50 kph, have reduced their hindwings very considerably in size and latched them on to the long narrow forewings with a curved bristle. Beetles use their forewings for a different purpose altogether. These creatures are the heavy armoured tanks of the insect world and they spend a great deal of their time on the ground, barging their way through the vegetable litter, scrabbling in the soil or gnawing into wood. Such activities could easily damage delicate wings. The beetles protect theirs by turning the front pair into stiff thick covers which fit neatly over the top of the abdomen. The wings are stowed beneath, carefully and ingeniously folded. The wing veins have sprung joints in them. When the wing covers are lifted, the joints unlock and the wings spring open. As the beetle lumbers into the air, the stiff wing covers are usually held out to the side, a posture that inevitably hampers efficient flight. Flower beetles, however, have managed to deal with this problem. They have notches at the sides of the wing covers near the hinges so that the covers can be replaced over the abdomen, leaving the wings extended and beating.

The most accomplished aeronauts of all are the flies. They use only their forewings for flight. The hindwings are reduced to tiny knobs. All flies possess these little structures but they are particularly noticeable in the crane flies, the daddy-long-legs, in which the knobs are placed on the ends of stalks so that they look like the heads of drumsticks. When the fly is in the air, these organs which are jointed to the thorax in the same way as wings, oscillate up and down a hundred or more times a second. They act partly as stabilisers, like gyroscopes, and partly as sense organs presumably telling the fly of the attitude of its body in the air and the direction in which it is moving. Information about its speed comes from its antennae, which vibrate as the air flows over them.

Flies are capable of beating their wings at speeds up to an astonishing 1,000 beats a second. Some flies no longer use muscles directly attached to the bases of the wings. Instead they vibrate the whole thorax, a cylinder constructed of strong pliable chitin, making it click in and out like a bulging metal tin. The thorax is coupled to the wings by an ingenious structure at the wing base, and its contractions cause them to beat up and down.

The insects were the first creatures to colonise the air, and for over a 100 million years it was theirs alone. But their lives were not without hazards. Their ancient adversaries, the spiders, never developed wings, but they did not allow their insect prey to escape totally. They set traps of silk across the flyways between the branches and so continued to take toll of the insect population.

Plants now began to turn the flying skills of the insects to their own advantage. Their reliance on the wind for the distribution of their reproductive cells was always haphazard and expensive in biological terms. Spores do not require fertilisation and they will develop wherever they fall, provided the ground is sufficiently moist and fertile. Even so, the vast majority of them, from such a plant as a fern, fail to find the right conditions and die. The chances of survival for a wind-blown pollen grain are very much smaller still, for their requirements are even more precise and restricted. They can only develop and become effective if they happen to land on a female cone. So the pine tree has to produce pollen in gigantic quantities. A single small male cone produces several million grains, and if you tap one in spring, they fall out in such numbers that they create a golden cloud. A whole pine forest produces so much pollen that ponds become covered with curds of it – and all of it wasted.

Insects could provide a much more efficient transport system. If properly encouraged, they could carry the small amount of pollen necessary for fertilisation and place it on the exact spot in the female part of the plant where it was required. This courier service would be most economically operated if both pollen and egg were placed close together on the plant. The insects would then be able to make both deliveries and collections during the same call. And so developed the flower.

Some of the earliest and simplest of these marvellous devices so far identified are those produced by the magnolias. They appeared about a 100 million years ago. The eggs are clustered in the centre, each protected by a green coat with a receptive spike on the top called a stigma, on which the pollen must be placed if the eggs are to be fertilised. Grouped around the eggs are many stamens producing pollen. In order to bring these organs to the notice of the insects, the whole structure is surrounded by brightly coloured modified leaves, the petals.

Beetles had fed on the pollen of cycads and they were among the first to transfer their attentions to the early flowers such as those of magnolias and waterlilies. As they moved from one to another, they collected meals of pollen and paid for them by becoming covered in excess pollen which they involuntarily delivered to the next flower they visited.

One danger of having both eggs and pollen in the same structure is that the plant may pollinate itself, thereby preventing cross-fertilisation, the very purpose of all these complexities. This possibility is avoided in the magnolia, as in many plants, by having eggs and pollen that develop at different times. Magnolia stigmas will accept pollen as soon as the flower opens. Its own stamens, however, do not produce their pollen until later, by which time its eggs are likely to have been cross-fertilised by exploring insects.

The appearance of flowers transformed the face of the world. The green forest now flared with colour as the plants advertised the delights and rewards they had on offer. The first flowers were open to all that cared to alight on them. No specialised organs were required in order to reach the centre of the magnolia flower or the waterlily, no particular skill was needed to gather the pollen from the loaded stamens. Such blooms attracted several kinds of insects – bees as well as beetles. But a variety of visitors is not an unmitigated advantage, for they themselves are also likely to visit several kinds of unspecialised flowers. Pollen of one species deposited in flowers of another is pollen wasted. So throughout the evolution of the flowering plants, there has been a tendency for particular flowers and particular insects to develop together, each catering specifically for the other's requirements and tastes.

Right from the times of the giant horsetails and ferns, insects had been accustomed to visiting the tops of trees to gather spores as food. Pollen was an almost identical diet and it still remains a most important prize. Bees collect it in capacious baskets on their thighs and take it back to their hives for immediate consumption or for turning into pollen bread which is an essential food for their developing young. Some plants, among them species of myrtle, produce two kinds of pollen, one that fertilises their flowers, and another of a particularly tasty kind that has no value except as food.

Other flowers developed a completely new bribe, nectar. The only purpose of this sweet liquid is to please insects so greatly that they devote all possible time during the flowering season to collecting it. With this the plants recruited a whole new regiment of messengers, particularly bees, flies and butterflies.

These prizes of pollen and nectar have to be advertised. The bright colours of flowers make them conspicuous from considerable distances. As the insect approaches, it is provided with markings on the petals which indicate the exact placing of the rewards they seek. Some flowers intensify their colours towards the centre or introduce another shade altogether – as do forget-me-nots, hollyhocks, bindweed. Others are marked with lines and spots like an airfield to show the insect where to land and in which direction to taxi – foxgloves, violets and rhododendrons. There are more of these signals on flowers than we may realise. Many insects can perceive colours of the spectrum that are invisible to us. If we photograph what seem to be plain flowers in ultraviolet light, we can see many more such markings on the petals.

Scent is also a major lure. In most cases, the perfumes that insects find attractive, such as those produced by lavender, roses, and honeysuckle, please us as well. But this is not always the case. Some flies are attracted to rotting flesh as a food for themselves and their maggots. Flowers that enlist them as pollinators must cater for their tastes and produce a similar smell, and they often do so with an accuracy and pungency far beyond the endurance of the human nose. The maggot-bearing *Stapelia* from southern Africa has flowers that reek dreadfully of carrion but it also reinforces its appeal to flies with wrinkled brown petals covered with hairs that look like the decaying skin of a dead animal. To complete the illusion, the plant generates heat that mimics the warmth produced by corruption. The whole effect is so convincing that flies transporting *Stapelia*'s pollen not only visit flower after flower, but even complete the activity for which they visit real carrion – laying their eggs on the flower just as they do in a carcass. When these hatch, the maggots find that they are not provided with a meal of rotting meat but only an inedible petal. They die from starvation, but the *Stapelia* has been fertilised.

Perhaps the most bizarre imitations of all are those of some orchids that attract insects by sexual impersonation. One produces a flower that closely resembles the form of a female wasp complete with eyes, antennae and wings and even gives off the odour of a female wasp in mating condition. Male wasps, deceived, attempt to copulate with it. As they do so, they deposit a load of pollen within the orchid flower and immediately afterwards receive a fresh batch to carry to the next false female. The extent of this mimicry can be far greater than mere physical resemblance. The orchid's flowers are covered with waxes that correspond to an extraordinary degree to the sex-specific pheromones that cover the female wasp and which are just as attractive. These orchids produce no nectar. The reward they provide for their insect pollinators is not sex, but its illusion.

Sometimes insects are disinclined to collect pollen, preferring nectar, and will bypass the plant's strategies and become nectar thieves, biting their way through the flowers from the outside and inserting their proboscis into the nectar source without getting covered in pollen. Then the flowers have to have devices to force their pollen on the insect. Some blooms have become obstacle courses during which their visitors are pummelled by stamens and bombarded with pollen before they are able to leave.

Broom flowers are so constructed that if, for example a bee, lands, the stamens, packed under tension inside a sealed capsule of petals, shoot out and strike the underside of the bee, covering its furry abdomen with pollen. The bucket orchid from Central America drugs its visitors. Bees clamber into its throat and sip a nectar so intoxicating that after they have taken only a little they begin to stagger about. The surface of the flower is particularly slippery. The bees lose their foothold and fall into a small bucket of liquid. The only way out of this is up a spout. As the inebriated insect totters up, it has to wriggle beneath an overhanging rod which showers it with pollen.

Sometimes plant and insect become totally dependent one upon the other. The yucca grows in Central America. It has a rosette of spear-shaped leaves from the centre of which rises a mast bearing cream-coloured flowers. These attract a small moth with a specially curved proboscis that enables it to gather pollen from the yucca stamens. It moulds the pollen into a ball and then carries it off to another yucca flower. First it goes to the bottom of the flower, pierces the base of the ovary with its ovipositor and lays several eggs on some of the ovules that lie within. Then it climbs back up to the top of the stigma rising from the ovary and rams the pollen ball into the top. The plant has now been fertilised and in due course all the ovules in the chamber at the base will swell into seeds. Those that carry the moth's eggs will grow particularly large and be eaten by the young caterpillars. The rest will propagate the yucca. If the moth were to become extinct, the yuccas would never set seed. If the yuccas disappeared, the moth's caterpillars could not develop. Each species is inextricably in the debt of the other.

One further debt is clear. Flowers, exquisitely perfumed and graced with a multitude of colours and shapes, bloomed long before humans appeared on the earth. They evolved in order to appeal not to us but to insects. Had butterflies been colour-blind and bees without a delicate sense of smell, we would have been denied some of the greatest delights that the natural world has to offer.

FOUR

The Swarming Hordes

B y any standards, the insect body must be reckoned the most successful of all the solutions to the problems of living on the surface of the earth. Insects swarm in deserts as well as forests; they swim below water and crawl in deep caves in perpetual darkness. They fly over the high peaks of the Himalayas and exist in surprising numbers on the permanent icecaps of the Poles. One fly makes its home in pools of crude oil welling up from the ground; another lives in steaming-hot volcanic springs. Some deliberately seek high concentrations of brine and others regularly withstand being frozen solid. They excavate homes for themselves in the skins of animals and burrow long winding tunnels within the thickness of a leaf.

The number of individual insects in the world seems beyond any computation, but someone has made the attempt and concluded that, at any one time, there must be something of the order of 10 billion billion. Put another way, for every human being alive, there are over a billion insects – and together these insects would weigh perhaps 70 times as much as the average human being.

There are thought to be about four times as many species of insect as of all other kinds of living organisms put together. So far, we have described and named about 900,000 of them and there are certainly three or four times as many still unnamed, and perhaps even more awaiting the attentions of anyone who has the time, patience and knowledge to sit down and make a systematic review of them.

Yet all these different forms are variations of one basic anatomical pattern: a body divided into three distinct parts – a head bearing the mouth and most of the sense organs; a thorax filled almost entirely with muscles to operate the three pairs of legs beneath and, usually, one or two pairs of wings above; and an abdomen carrying the organs needed for digestion and reproduction. All three sections are enclosed within an external skeleton made primarily of chitin. This brown fibrous material, as we have seen, was first developed over 550 million years ago by the early segmented creatures, the crustaceans and probably the trilobites. Chemically, it is similar to cellulose and in its pure form it is flexible and permeable. The insects, however, cover it with a protein called sclerotin that makes it become very hard. This produces the heavy inflexible armour of the beetles, and mouthparts sharp and tough enough to gnaw through timber and even cut metals like copper and silver.

The chitinous external skeleton is particularly responsive to the demands of evolution. Its surface can be sculpted without affecting the anatomy beneath. Its proportions can be varied to take on new shapes. Thus the chewing mouthparts possessed by the early cockroach-like insects have been turned by their descendants into siphons and stilettos, saws and chisels, and probes that when unreeled are as long as the whole body. Legs have become elongated into catapults that can propel an insect two hundred times its own body length, broad oars to row it through water or thin hair-tipped stilts with a wide stride that enables their owners to walk on the surface of pools. Many limbs carry special tools moulded from chitin – pouches for holding pollen, combs for cleaning a compound eye, spikes to act as grappling irons and notches with which to fiddle a song.

An external skeleton, however, is also an unexpandable prison. The trilobites in the ancient seas escaped its restrictions by moulting. That is still the insects' solution. The process may sound wasteful, but they conduct it with great economy. A new chitinous shell, much wrinkled and compressed, forms beneath the old one. A layer of liquid separates the two and this absorbs the chitin from the old skeleton, leaving the hard sclerotinised parts connected by the thinnest of tissues. The chitin-rich liquid is then absorbed through the still permeable new skeleton back into the insect's body. The old plates split apart, usually along a line running down the back, and the insect hauls itself out. As it does so, its liberated body begins to swell, filling out the folds in the new skin. In a short time, the chitin hardens and becomes strengthened by new deposits of sclerotin.

Primitive relatives of the insects like the bristletails and the springtails do not change their shape very much as they grow. They merely moult as they increase in size. Even after they have begun to breed they may continue to moult. The ancient winged insects – cockroaches, cicadas and crickets – also grow in a similar way, their early forms closely resembling the adults except that they lack wings. Even when these insects adopt a very different existence for the first part of their lives, they do not change their form very radically. The larvae of the cicadas that sit shrilling on trees spend their lives below ground sucking sap from roots. Larval dragonflies hunt on the bottom of ponds, grabbing worms and other small creatures with long protrudable mouthparts. Yet in both the cicada and the dragonfly the image of the adult is discernible.

More advanced insects, however, undergo such wholesale changes that there is no possible way of linking the larva to the adult except by watching the creature make the change. Maggots turn into flies, grubs into beetles, and caterpillars into butterflies.

The job of a grub, a maggot or a caterpillar is simply to eat. Its body is dedicated to this one purpose. Since it will not breed in this form, it has no sexual equipment; since it has no cause to attract a mate, it needs no mechanisms to send out call signals whether by sight, smell or sound, nor any sense organs to receive such messages; and as its parents have gone to considerable trouble to ensure that when it hatches it is surrounded by great quantities of the particular food it requires, it needs no wings. Its one essential tool is a pair of efficient jaws. Behind these it requires little more than a bag. In order that this may swell easily to accommodate its rapidly accumulating tissues, this simple body is not burdened with a heavy sclerotised skeleton but enclosed in a thin and, to some degree, stretchable cuticle. When this can expand no further, it splits and is rolled off, like a nylon stocking from a leg.

With no shell and thus no firm base on which to attach a muscle and nothing rigid to serve as a lever, these larvae are indifferent movers. They cannot hop, skip or jump. Indeed, they can barely manage even to run, for they have only soft ballooning tubes to serve as legs. These stumpy limbs however are quite efficient enough to move the eating machines that are their owners from one mouthful to another.

The lack of shell leaves the larvae unprotected. This is of little consequence to grubs and maggots, for they hold their interminable feasts out of sight of the rest of the world, gobbling their way around the heart of an apple or gnawing tunnels in wood, shielded by what they are eating. But caterpillars, most of which banquet out in the open, must look to their defences.

They are matchless as camouflage artists. Those of geometer moths are coloured and patterned to look like twigs, and when they hold themselves with one end in the air at exactly the same angle to a stem as other twigs springing from it, they are virtually impossible to detect. A swallowtail caterpillar, sitting on a leaf, is certainly conspicuous, being green with irregular flecks of white, but it is seldom noticed for it looks like a bird dropping. If disguises are penetrated, many caterpillars have a second line of defence. The pussmoth caterpillar browses head-down on leaves. Its body colour is exactly that of its food plant, but if an intruder shakes the branch and alarms it, the caterpillar suddenly lifts its head from its meal, exposing a scarlet face. Simultaneously it protrudes a pair of blood-red filaments from its tail and squirts formic acid. Another moth caterpillar in South America can be even more alarming. It has a large round mark on either side of its head, and, when agitated, weaves its front end from side to side, making itself look unnervingly like a wide-eyed snake.

Some caterpillars have made themselves unpleasant to eat. They are covered with poisonous hairs or have within their bodies a particularly acrid-tasting substance. It pays these creatures to be very conspicuous indeed. The hairy ones are moustachioed and bewhiskered in the most flamboyant way, the unpleasant-tasting ones have skins brilliantly coloured in reds, yellows, blacks and purples – all warnings to potential hunters that these morsels, for one reason or another, are not worth eating.

There are also some insects that, as larvae or adults, are actually innocuous but seldom eaten for they have taken a rather complicated gamble by copying the colours of poisonous caterpillars or stinging adults to delude aggressors into giving them as wide a berth as the creatures they mimic. The garden hoverfly, with its black-and-yellow-striped abdomen, deceives not only predators but many human beings into mistaking it for its model, a wasp. In harmless flies, such mimicry can be so precise that they tremble their abdomen as a bee does or produce the smell of a wasp.

Many insects spend nearly all their lives as such larvae, growing bigger and building up their stores of food. Beetle grubs may spend seven years boring through wood and extracting nutriment from that most indigestible of materials, cellulose. Caterpillars munch for months, packing away their favourite leaves before the season finishes. But sooner or later, they all reach their full size and the end of the allotted span of their larval lives.

Now comes the first of two highly dramatic transformations. It is a change some make in private. Only the larvae of insects have silk glands. They have used them already to build communal tents, to extrude lifelines guiding them over plants, or as ropes to let themselves down from one twig to another. Now, however, many spin silk to conceal themselves from the world. The silk moth caterpillar surrounds itself with a fuzzy bundle of threads, the moon moth constructs a cocoon with a silvery metallic sheen, the ermine moth builds an elegant casket of lacy net. Many butterfly larvae produce no covering at all. They simply spin a silken sling and use it to attach themselves to a twig.

As soon as they are settled, they discard their caterpillar costumes. Their skin splits and rolls down, revealing a smooth, brown hard-shelled object, the pupa. The only movement it makes is an occasional twitch of its pointed tip. It has spiracles along its side through which it can breathe, but it neither feeds nor excretes. Its life seems to have been suspended. Internally, however, the most profound changes are taking place. Much of the body of the larva is being dismembered and reassembled.

When the larva first began to develop from the egg, its cells were segregated into two groups. Some stopped dividing after a few hours and remained generalised in form and in dense clusters. The rest continued to build the body of the caterpillar. After it hatched and had begun to feed, its body cells divided no more. Instead they simply enlarged until, by the time that the caterpillar was full grown, they were vastly distended and many thousand times bigger than their original size. All this time, the other cell clusters remained tiny and inactive. But towards the end of the caterpillar phase, their moment comes. The dormant cell clusters suddenly begin to divide rapidly, slowly building a new body with a completely different form. Naturalists in the seventeenth century who dissected caterpillars on the verge of pupation were able to recognise the soft outlines of the wings, head and legs of the future butterfly, which convinced them that the caterpillar and the butterfly are one and the same organism rather than that the butterfly arose out of the death and decay of the caterpillar, as had previously been thought. These features also become visible on the outside of the brown pupa when it hardens, like those of a mummy, shadowy beneath its wrappings. Indeed, the name 'pupa' derives from the Latin for doll.

The exact process of metamorphosis is still not fully understood. Different groups of insects show different degrees of change. The most radical alteration comes in the flies where the future adult organs exist merely as patches of skin within the maggot's body until after the pupa has formed. But even here, the essential parts of the maggot, including its brain and central nervous system, are preserved and used as scaffolding to build the fly. There are even tantalising hints that those insects that undergo metamorphosis may remember as adults things they learned as larvae.

The actual emergence usually takes place under cover of darkness. A butterfly pupa, hanging from a twig, begins to shake. A head with two huge eyes and antennae pressed over its back pushes through the pupa at one end. Legs come free and begin clawing frantically in the air. Slowly and laboriously, with frequent pauses to gather strength, the insect hauls itself out. The thorax emerges and there on its back are two flat crumpled objects, its wings, wrinkled like the kernel of a walnut. The insect jerks itself free and hangs on the empty pupa case, its body trembling. Then, with convulsive shudders, it begins to pump blood into a network of veins within the baggy wings.

Slowly they expand. The blurred pattern on the outside of the wings enlarges and becomes focused. Blotches swell into miraculously detailed eye-spots. Within half an hour the wings are fully distended so that the two sides of the bag meet flat against one another, enclosing the veins between them. The veins themselves are still soft. If the tip of one of them were damaged now, it would drip blood. But gradually the blood is drawn back into the body and the veins harden into rigid struts that will give the wing its strength. All this time, the wings have been held together like the leaves of a book. Now, as they dry and become rigid, the insect slowly moves them apart to show the world for the first time the unblemished perfection of its shimmering colours and awaits the dawn of its first day.

The insect can now spend the calories that it so assiduously collected and stored when it was a larva. For the adult, feeding is of secondary importance. Some sip nectar during their brief lives to renew their energies and to provide sustenance for egg production, but none need to feed in order to build their bodies; their growth has come to an end. Mayflies and some moths do not even have mouthparts. The urgency of their lives now is to find a mate.

Butterflies do so by displaying their wings, the marvellous intricate patterns of which are statements of identity, both of species and sometimes sex, so that individuals may recognise those with whom mating can be fertile. Unlike their larvae, butterflies have excellent compound eyes – the male usually has even bigger ones than the female, for it is generally he that does the searching. Since their eyesight is sensitive to parts of the spectrum that are invisible to us, butterflies' wings, like flowers, have even more complex patterns than our ultraviolet-blind eyes can see. Colours and designs created by tiny scales, overlapping like tiles on a roof, come from pigments or, more often, the effects of microscopic structures which split the light falling on them and reflect back only a part of it. Drop a spot of highly volatile liquid on such a wing and the colours disappear as the liquid occludes the physical structure, only to reappear as it evaporates and the light is splintered once more.

These dazzling wings, iridescent and downy, trailing pennants and variegated with transparent windows, veined, fringed and spotted with the loveliest of colours, are the most elaborate visual summons in the whole insect world. Other insects use other media and produce equally complex and powerful signals to call across distance. The cicadas, crickets and grasshoppers rely on sound. Most insects are deaf, so these groups have had to develop not only voices but ears. Cicadas have circular eardrums on either side of the thorax. Grasshoppers listen with their legs. They have two slits on the first pair of thighs which lead to deep pockets. The common wall between these forms a membrane that is the equivalent of an eardrum. The angle at which sound strikes the slits greatly affects the strength in which it reaches the drum, so the grasshopper, by waving its legs in the air, can discover the direction from which a call is coming.

Some grasshoppers produce their whirring trills by sawing the notched edge of their hindlegs against a prominent strengthened vein of the wing. Cicadas, the loudest of insect singers, have a much more complicated apparatus. Their abdomen contains two chambers, one on each side. The inner wall of each chamber is stiff, and when it is moved in or out, it makes a click, as the lid of a tin may do. In the abdomen behind there is a muscle which can pull the wall back and forth up to 600 times a second. The noise this produces is greatly amplified, for most of the abdomen behind the vibrating plate is also hollow, and two large rectangular sections of the abdominal wall are stiffened to form resonators. These are covered by flaps projecting from the lower edge of the thorax that can be opened or closed so as to increase or dampen the sound like shutters on an organ. Each species produces its own characteristic call. Some sound like a mechanical saw hitting a nail, some like a knife being ground on a wheel or fat dropping on an overheated plate. So loud are these calls that a single insect can be heard half a kilometre away and a chorus of them can set a whole forest ringing and echoing.

There is much more detail in these penetrating songs than our ears can detect. We cannot hear a break between sounds of less than one tenth of a second. Cicadas are able to distinguish intervals of one hundredth of a second. When they sing, they vary the frequency of individual clicks from, for example, 200 a second to 500 a second and do so in a regular rhythmic way. By such changes and rhythms, which are totally inaudible to us, an individual can identify the call of its own species; a male can avoid the territory of another singing male, and a female fly towards it.

Mosquitoes also use sound as a mating call, but they produce and receive it in a way that is all their own. The female, beating her wings as fast as 500 times a second, creates the high-pitched hum that can be so unsettling as you lie in camp trying to go to sleep without a mosquito net. The male is able to detect this sound with an eardrum at the base of his antennae, which vibrate in sympathy with this one frequency, and so fly towards her.

Other insects attract their mates by exploiting the third of the senses, smell. The females of some moths produce an odour that the males can detect with large feathery antennae. So sensitive are these organs and so charac-teristic and powerful is the scent, that a female has been known to summon a male from eleven kilometres away. At such a distance there may be as little as one molecule of scent in a cubic metre of air, yet it is sufficient to cause the male to fly in pursuit of its source. He needs both antennae to do this. With only one, he cannot establish direction, but with two he can judge on which side the scent is stronger and so fly steadily towards it. A female emperor moth, in a cage in a wood, transmitting a perfume undetectable to our nostrils, has attracted over 100 huge males from the surrounding countryside within three hours.

So, by sight, sound and smell, the adult insects attract their mates. Male grasps female sometimes only briefly, sometimes for several hours. The couple may even fly through the air awkwardly, in tandem. Then the female lays her fertilised eggs and provisions them. Butterflies seek out the one plant whose leaves provide the only food their caterpillars will eat; beetles bury pellets of dung and lay their eggs within them; flies feverishly deposit their eggs within carrion; and solitary wasps catch spiders, paralyse them with a sting and stack them around their eggs so that the young larvae will have fresh meat awaiting them. The female ichneumon wasp has an ovipositor like a dagger with which she drills a hole in wood at the exact point where she has detected a beetle grub lying beneath. Her ovipositor pierces it and deposits an egg in its soft body. When it hatches, her larva will then eat the grub alive. And so the whole process of egg-larva- pupa-adult begins once more.

The insect body has produced an almost infinite variety of forms. In one characteristic only does there seem to be a limitation – size. The largest living insects today are not longer than about 30 centimetres – the wingspan of exceptional specimens of the atlas moth, and the length of the largest of the stick insects. The biggest of beetles, the hercules, reaches the same sort of size and weighs as much as 100 grams. But that is only the size of a mouse. Why are there not beetles as big as badgers and moths as large as hawks?

The restricting factor is their breathing technique. Like their close relatives, the early millipedes, the insects rely on tracheae, the system of tubes opening to the outside by a line of spiracles along the flank and running to every part of the body. They work by gaseous diffusion. Oxygen in the air that fills the tracheae is absorbed through the wall at the extremities. Similarly, carbon dioxide is expelled from the tissues and diffuses away. The system works excellently over short distances, but as the length of the tube increases, so it becomes more and more inefficient. Some insects are able to improve the circulation of air by inflating and deflating their abdomens with a muscular pumping action. The tiny tracheae, which are strengthened with rings in their walls, do not flatten but shorten and expand like concertinas. A few insects have tracheae that swell into thin-walled balloons which are depressed and expanded as the abdomen pumps up and down. But even with these refinements the system is ineffective above a certain size; the gigantic cockroaches and murderous man-hunting wasps of nightmares are physiological impossibilities.

But the insects have, in another way, transcended even the limitation of scale. All over the tropics stand hills of bare hardened mud made by termites. In some parts they are grouped in swarms hundreds strong, as thick as herds of grazing antelope. The comparison is not entirely fanciful. A single hill contains a colony of several million insects. They are not just creatures that have elected to live together in one communal dwelling, like human beings in some gigantic tower block. For one thing, they are all one family, the offspring of a single pair of adults. For another, all of them are incomplete creatures, incapable of independent life. The workers, scurrying along the tracks through the undergrowth, are mostly blind and all are sterile. The soldiers that stand guard beside the entrances to the colonies and rush to defend any breach in the walls, are armed with jaws so huge that they can no longer gather food for themselves and have to be fed by the workers. At the centre of the colony lies the queen. She is imprisoned within massive earthen walls from which she can never escape, for her body is far too big to get through the passages that lead to it. Her abdomen is swollen into a white heaving sausage, 12 centimetres long, from which she produces eggs at the almost unbelievable rate of 30,000 a day. She too would die if she were unattended. Teams of workers deliver food to the front end of her body and collect eggs from the back. The only sexually active male, the wasp-sized king, stays alongside her and he too is fed by the workers.

The link that binds all these individuals together into one coordinated superorganism is a highly effective system of communication. Soldier termites sound an alarm by beating their large hard heads on passage walls. Workers, having a new source of food, leave a scent trail which their blind fellows can easily follow. But the most pervasive and important mechanism is based on chemical substances called pheromones. This circulates instructions throughout the colony with great speed. All the members of the colony continually exchange food and saliva with one another. Workers pass it from mouth to mouth or gather one another's excrement in order to reprocess the partially digested food and extract the last particle of nutriment from it. They in turn feed both the larvae and the soldiers. They also attend the queen lying in her chamber, constantly licking her rippling flanks and collecting drops of liquid from her anus. In the course of this they gather the pheromones that she produces and circulate them quickly throughout the colony. The young larvae, hatching from the queen's eggs, are potentially of both sexes, but the queen's pheromones with which they are fed by the workers inhibit their development and they remain sterile, wingless and blind. The soldiers too produce a pheromone, contributing to the mix of chemical messages circulating in the colony and in a similar way preventing the development of any of the larvae into soldiers.

At certain times of the year, the blend of pheromones produced by the queen or the response of the larvae to it, shifts slightly. The mixture loses its power to inhibit and the dark corridors of the colony are filled with rustling hordes of young winged adults. In some species, the workers open special slits in the sides of the mound and build take-off ramps in front of them. These exits are guarded by the soldiers. Then, just after the beginning of the rains, the soldiers stand aside and flying termites pour out of the clefts and swirl into the sky like smoke.

The occasion is a bonanza for the animals of the bush. Frogs and reptiles gather beside the exits, snapping at the insects as they flock out on to the ramps. As the exodus proceeds, the sky becomes filled with birds wheeling back and forth. The termites seldom go far. They come down on to the ground and immediately their wings break off close to the thorax. They have served their function. Now male chases female across the ground. Those few that escape being eaten form pairs and go off together to find a nest site in a crevice in the ground or a crack in a tree. There they construct a small royal cell. Within it, they copulate and lay eggs. The first larvae to hatch have to be fed by their parents, but once they are big enough to forage for food and build walls of mud, the royal couple devote themselves entirely to the production of eggs and the colony is founded.

Termites are closely related to those ancient insects, the cockroaches. Like them, their bodies do not have a waist and the young larvae are markedly similar to the adult winged form. They grow by a series of moults but never pass through a pupal stage or undergo transformation. Like cockroaches too, the termites feed almost entirely on vegetable matter. There are some 2,000 different species of them. Twigs, leaves and grass are standard fare. Some specialise in eating timber, boring away inside posts and logs, reducing them to hollow shells that can collapse at the touch of a finger.

Termites construct some of the greatest of all insect buildings. A termite fortress, walled, buttressed and castellated, may contain ten tonnes of mud and stand three or four times as tall as a human. Several million inhabitants, busily running their errands within, can cause overheating and produce a foul oxygen-poor atmosphere so ventilation is of the greatest importance. Around the margins of the hill, the termites construct tall, thin-walled chimneys which stand out from the sides like ribs. No insects live inside these huge smooth-walled ducts. Their only function is ventilation. As the sun warms their walls, the air inside becomes hotter than that in the centre of the nest. It rises, drawing exhausted air from the central galleries and the deeper parts of the hill, creating a circulation. The thin, external walls of the chimneys are porous and so oxygen from the outside atmosphere diffuses in. The air, thus refreshed, rises to the top of the nest and then circulates back down other passageways. In very hot weather, the workers descend tunnels that go deep into the ground to the water table. Each returns carrying a crop full of water with which it wets the walls of the main part of the nest. The heat evaporates the water and this also lowers the temperature. By such devices, the worker termites manage to keep a very even temperature inside the nest.

In Australia, the compass termites build castles in the shape of huge flat chisel blades, always with their long axis pointing north and south. Such a shape exposes the minimum possible area to the ferocious midday sun but catches the maximum of the feebler rays in the early morning and evening when, especially in the cold season, the termites are grateful for warmth. In West Africa and other areas where there is heavy rain, the colonies build nests like mushrooms with flat roofs which shed the water. Termitologists have made great advances in working out how the pheromone communication system controls and coordinates a colony's activities, but no one has yet explained how the millions of blind workers, each carrying a tiny pellet of mud, manage between them to construct such ingeniously shaped, efficient and large-scale buildings.

One other group of insects has taken to the colonial life on a scale that is comparable to the termites; those with narrow waists, two pairs of transparent wings and powerful stings, the wasps, bees and ants. Wasps still show the stages by which colonialism may have developed. Some hunting wasps live entirely solitary lives. The female, after mating, builds her own mud cells, lays an egg in each, provisions it with a collection of paralysed spiders and then abandons it. In other species, she stays beside the nest and, when the young hatch, brings food to them day after day. In yet others, closely related females build their individual nests close to one another but, after a few weeks, some abandon their own constructions and join others in building theirs. Eventually, one female becomes dominant and lays all the eggs while the others concentrate on building cells and collecting food for her.

The honeybees have taken this basic arrangement and elaborated it to an extreme degree so that they live in colonies of many thousands. The single queen stays on the comb, laying eggs in the cells that have been built by the workers to receive them. Once again, the community, like that of the termites, is bound together by a system of chemical messages, the pheromones, perpetually circulating within the hive, which inform all the inhabitants of the state of the population and of the absence or presence of the queen. If the queen dies or if her pheromones indicate a flagging level of egg production, the workers will begin to rear a new queen and will also start to produce their own eggs which are inevitably unfertilised, for they cannot mate. These nonetheless hatch and produce males – drones. They do nothing useful in the hive. They are simply flying packets of sperm able to spread the workers' genes to other hives should no queen develop before the colony dies.

Bees have an astonishing way of communicating with one another that is unique among colonial insects. Flying through the air to find food, they cannot leave scent trails behind them to guide other members of the colony as earth-bound termites do. Instead, they dance. When a worker bee arrives back in the hive after visiting a newly opened nectar-laden flower, it performs a special dance on the landing platform in front of the entrance to the colony. First it scurries round in a circle, then it bisects it, emphasising the importance of this last movement by waggling its abdomen and making a particularly excited kind of buzzing. Its track points directly to the source of food. Workers observing it and about to start on their own foraging, immediately fly off in the direction indicated. Then the dancer goes into the hive to dance again. The farther it goes from the entrance before it dances, the farther away is the flower that it has discovered. The combs of the nest, in both wild and domestic colonies, are vertical so now the waggle-steps cannot point directly to the food source. Instead they refer to the sun. If the bee crosses the circle vertically, then the target is in line with the sun. If it is, say, 20° to the right, then the dance will be 20° to the right of the vertical. The workers surrounding the dancer watch it closely, remember the message and fly away to find the flower. When they return with the honey, they too will perform a dance so that, in a very short time, most of the worker force in the hive is actively gathering honey from the new source.

This much was worked out by scientists from the 1940s onwards but we still do not fully understand how the waggle-dance works. After all, it is dark in the hive where the the dance takes place, so the workers cannot see the angle at which the dance is performed. They must be able to sense it through the sound or smell produced by the dancer. Ingenious experiments with tiny robot bees have shown that buzzing is a necessary component of the dance. Without it the workers are much less impressed.

The most complex and sophisticated forms of sociality in the insect world are those created by the relations of the wasps and bees, the ants. Some live within plants, stimulating the tissues of their hosts to provide them with custom-built homes by growing special galls, hollow stems or thorns with swollen bases. The leaf-cutting ants of South America build vast underground nests and set off from them, day and night, in long columns to demolish trees, removing every shoot, leaf and stem, section by tiny section, and transporting them all back to their underground chambers. They do not eat this material but chew it up to form a compost on which they cultivate a fungus. Its tiny white fruiting bodies then provide them with their food.

Tree ants in Southeast Asia construct nests by sewing leaves together. A party of workers haul two leaf edges together, gripping one with their jaws and the other with their feet. Other workers on the inside begin the work of sewing them together. No adult insect can produce silk, so these ants bring their own larvae to the site, holding them between their jaws and giving them little squeezes so that the larvae are stimulated to produce their silk. The builders then move these living tubes of glue in their jaws back and forth across the leaf junction until the two edges are joined by a silken fabric. In Australia, the honey pot ants collect nectar and feed it to workers of a special caste until their abdomens are distended to the size of peas and their skins stretched so thin that they are quite transparent. The workers then hang them up by their forelegs in underground galleries, like living storage jars.

Most ants, however, are carnivorous. Many prey upon termites, raiding the great mounds and doing battle with the soldiers. If they win, they devour the defenceless workers and larvae. Others, in one of the most astounding forms of social behaviour, make slaves of a different kind of ant. They raid a nest of their victims, which are generally of a closely related species, collect the pupae and carry them back to their own colony. When these hatch, the young ants emerge into a world in which they are surrounded by their sisters – or so they think, because the pheromones on the slave-makers are similar to those of their own kind. Unable to detect that these are not relatives – indeed, they are not even the same species – they serve their captors, collecting food and rearing the brood just as they would have done in their own nest if they had not been captured and translocated. In obligatory slave-making species, which have become in effect social parasites, the slaves even feed their captors, for the slave-makers have such large jaws that they cannot feed themselves.

The most terrifying ants of all are those that do not make nests but range through the countryside seeking prey. In South America they are known as army ants, in Africa as drivers. They march in columns so long they may take several hours to pass one spot. At the head, the soldiers fan out to forage. Behind comes a column of workers, scurrying along a dozen or so abreast, many of them carrying larvae. Where the column crosses an exposed area its flanks are guarded by soldiers, armed with huge jaws and totally blind. They stand in rows, stiffly craning themselves upwards, jaws apart, ready to bite anything that interferes with them. When the hunters at the head of the column discover prey, they swarm all over it and butcher it. Grasshoppers, scorpions, lizards, young birds in their nests, anything that cannot get out of the way is attacked. Anyone in West Africa who tethers an animal or restricts its movement must pay regard to the possibility of an attack by one of these armies. I once made a large collection of snakes there. We had Gaboon vipers, puff adders, spitting cobras as well as non-venomous species like tree snakes and pythons. We kept them in a mud-walled hut and posted a guard to keep watch, armed with a can of paraffin. Only that, poured on the ground and set alight, will deflect a raid by the ants. In spite of all precautions, a column got into the hut through a hole in the wall at the back. By the time we had discovered what had happened, the ants were attacking the entire collection, swarming over the snakes within their gauze-covered boxes. Infuriated by the painful bites, the snakes were striking dementedly and uselessly at their tiny attackers. Every one had to be taken out and held down while we picked off the ants that were sinking their jaws between the scales. In spite of all we could do, several snakes died as a result of the ant bites.

Army ants march and forage for weeks, day after day. The larvae produce pheromones and these, circulating within the army, stimulate it to keep on the move. Eventually, the larvae begin to pupate and no longer exude their chemical messages. Then the army bivouacs. There may be as many as 150,000 individuals and they cluster in a vast ball between the roots of a tree or beneath an overhanging stone. Clinging to one another, they make a living nest from their bodies, complete with passageways along which the queen moves and chambers where the pupae are deposited. The queen's ovaries now begin to develop and she swells greatly. After about a week, she begins to lay. During the next few days, she may produce 25,000 eggs. They hatch very quickly and at the same time a new generation of workers and soldiers emerges from the stored pupae. These now begin to secrete their characteristic pheromone and once more the army, with its ranks swollen by these new recruits, marches off to war.

If a colony of termites can be compared to an antelope, then the disciplined aggressive columns of the army ants must be reckoned to be the equivalent of a beast of prey. Hungry for food, relentless in pursuit of it and capable

FIVE

The Conquest of the Waters

Among the sea anemones sticking limply to the rocks exposed at low tide, there are, almost everywhere in the world, lumps of jelly that are rather different. Anemones tend to dribble a little water from their centres if you press them. Tread on one of these others and a jet of water squirts up your leg. They are therefore called, not unreasonably, sea squirts. Underwater, the difference between them and the anemones is easy to see. The anemone has a flower-like cluster of tentacles around a single central opening; the sea squirt has no tentacles and two openings connected to each other by a U-shaped tube surrounded by a thick coat of jelly. Underwater and dilated, this dull bag becomes beautiful. One European species is nearly transparent, with trembling circlets of hazy blue around each opening and thin rings of muscle strengthening the inner tube so that the creature looks like the most delicate bubble of Venetian glass. The jacket of jelly of other species is opaque and coloured pink or gold. Some grow in clusters like grapes; some are larger, more elongated and solitary.

They are all filter feeders, drawing water in through one opening, passing it through a bag with slits in its wall, and then discharging it back into the sea through the other opening. Food particles, sticking to the wall of the bag, are swept down to its bottom by cilia and into a little gut which leads out of the bottom of the bag and curves round to join the exhalant tube.

Theirs is a simple structure and an unobtrusive life. But these creatures have very sophisticated relatives. Their most ancient forebears became the ancestors of the first of the backboned animals, the vertebrates. The evidence for this surprising claim is barely possible to distinguish in the adult sea squirt, but visible in its larva. This looks like a tiny tadpole. The globular front part contains the U-shaped tube and the beginnings of the gut. It swims by waggling its tail, which is stiffened by a thin rod, the notochord, which runs from the tip to the middle of the body. This, at least, is some suggestion of a backbone, but the larva does not keep it for very long. After a few days, the little creature glues its nose on a rock, loses its tail and settles down to a life of sedentary filtration.

The sea squirt larva is not the only filter feeder with such a significant rod in its back. Another somewhat larger sliver of flesh, the lancelet, also has one. This creature, shaped like a slim leaf about 6 centimetres long, lives half buried in the sand of the seafloor. Its front end projects above the surface and carries a little coronet of tentacles around the opening through which it sucks in water. It, too, has a very simple body. There is nothing that could reasonably be called a head; merely a small light-sensitive spot; no heart, only a number of pulsating arteries; no fins or limbs, only a slight dilation at the hind end like the flight feathers of an arrow. Even so, in this simple organism you can see the first hint of a fish. The flexible rod in its back which runs the entire length of its body carries transverse bands of muscles. When the creature contracts them rhythmically, a series of waves runs down its body. These push water backwards and in consequence the lancelet moves forward. It swims.

When assessing family connections, the anatomy of a larva is obviously as valid a piece of evidence as that of the adult. Indeed, it is usually even more significant, for many animals have the remarkable tendency to repeat during their individual development the stages through which their ancestors passed during evolutionary history. Larval termites look like those most primitive of insects, the bristletails; larval horseshoe crabs are visibly segmented and so reveal a similarity with the trilobites difficult to perceive in the adult; the free-swimming molluscan larva looks very like that of the segmented worms and thus suggests a link between the two groups. So it is not unreasonable to regard the similarity between the lancelet and the larval sea squirt as evidence of a relationship. But which form came first? Was it a creature like a sea squirt which gave rise to the more mobile lancelet-like form by producing descendants that abandoned the stationary condition and reproduced during the hitherto larval stage? Or was the lancelet shape the more ancient pattern from which animals like sea squirts developed by sticking their heads to rocks, losing their muscles and retreating into as undemanding a lifestyle as the seas can provide?

For many years, the first proposition was believed to be the case. Today comparative studies and above all genetic analysis of these organisms have shown that the second possibility is, in fact, the correct one. Confirmation has come from that remarkable treasury of early fossils in the Canadian Rockies, the Burgess Shales. There, lying among trilobites, brachiopods and bristle worms, in the mud of the seas of 550 million years ago that had yet to see a finned or backboned swimmer, lie the impressions of a creature very similar indeed to the living lancelet.

Another larva provides evidence for the next step in vertebrate history. The rivers of Europe and America contain animals that look like the lancelet, though they are somewhat larger, up to 20 centimetres long. They, too, live in holes in the mud and filter-feed. They are jawless, eye-less and without fins except for a fringe around the tail. For many years they were thought to be adult creatures and therefore given a special name and classified as obvious relatives of the lancelet. Then it was discovered that they are only the larvae of a very well-known animal for they eventually leave their holes, develop true eyes and long rippling fins along the back, grow to the size of an eel and become lampreys.

You might be excused for thinking, at first sight, that the lamprey is a true fish. But it is not. It has a kind of backbone, in the form of the flexible rod, but it does not have jaws. Its head ends in a large circular disc in the centre of which is a tongue covered with sharp spines. There are two small eyes with a single nostril between them leading to a blind sac, and on either side of the neck a row of gill slits. With the disc, the lamprey clamps itself on the flank of a fish, and with the tongue it rasps off the flesh, eating the fish alive. Lampreys and their wholly sea-living relatives, hagfish, are still common. Sometimes their population in American rivers reaches plague proportions. The swarming lampreys consume not only dead or sickly fish but set upon otherwise healthy ones. Their tiny eyes, rubbery sucking mouths and writhing bodies scarcely make them attractive from a human point of view. Nonetheless, they deserve notice and respect, for their forebears were once the most advanced and revolutionary creatures in the seas. Their abundant remains have now been found in the younger layers of the Burgess Shales, which are about 500 million years old, as well as in similar deposits in China.

These jawless proto-fish were mostly quite small, the size of large minnows, but they had eyes, a nose and arched structures that supported their gills. Some of them were armoured, the whole of the head and body encased in a suit of bony plates. At the front there were two eyes and a single central nostril, like that of a lamprey. From the back of the armoured box projected a muscular tail fringed by a fin. By beating this they could drive themselves through the water, but their heavy foreparts must have kept their heads low and close to the bottom. Although one or two species later developed simple flaps of skin behind their heads, many had nothing except their tail to help them steer or give precision to their movements. This may have meant that they found it difficult initially to swim much above the seafloor. Those upper waters still remained the domain of the jellyfish and other floating invertebrates. Without jaws, the proto-fish could not prey upon shelled molluscs, but they had evolved a new way of eating. They were able to exert a powerful suction through their simple circular mouths as they nuzzled their way across the seabed, sucking up mud, filtering out the edible particles and expelling the rest through the slits on either side of their throats.

The heavy deposits of bone in the head regions of some of these proto-fish make possible the most detailed investigation of their anatomy. By taking a series of slices through a fossilised skull, the shape of the cavities that contained the nerves and blood vessels can be charted. Such research has shown that one group of these creatures had a brain very like that of the living lamprey. It also had a balancing mechanism contrived from two arching tubes at right angles to one another in a vertical plane. The liquid within them, moving over their sensitive inside surface, must have enabled such a proto-fish to be aware of its posture in the water. Living lampreys have a very similar mechanism.

Some of these creatures now grew to a considerable size, 60 centimetres or so. Many were quite mobile, with suits of scales, and were probably able to make sallies into the waters well above the seafloor. None of them, however, could be described as skilled swimmers. The single median fins down the mid-line of their backs or undersides prevented them from spinning in the water and gave them a degree of stability, but none had paired lateral fins on their bodies.

So the situation remained for millions of years. During this immensity of time, the corals appeared and the segmented animals developed into forms that soon would leave the sea and establish a bridgehead on land. While they were doing so, important changes were taking place among the proto-fish. The slits in the sides of their throats had originated as filtering mechanisms, but they were walled with thin blood vessels so that they also served as gills. Now the pillars of flesh between them were stiffened with bony rods and the first pair of these bones, slowly over the millennia, in sucessive generations gradually hinged forwards. Muscles developed around them so that the front ends of the rods could be moved up and down. The creatures had acquired jaws. The bony scales in the skin that covered them grew larger and sharper and became teeth. No longer were the backboned creatures of the sea lowly suckers of mud and strainers of water. Now they could bite. Flaps of skin grew out of either side of the lower part of the body, helping to guide them through the water. These eventually became fins. Now they could swim. And so, for the first time, vertebrate hunters began to propel themselves with skill and accuracy through the waters of the sea.

It is possible to walk across the seabed of the period when that happened, 400 million years ago. In the flat desert land of a cattle station in northwestern Australia, close to a place called Gogo by the Aboriginal people, rises a line of strange steep-sided rocky bluffs, 300 metres high. Geologists, mapping the site, found it difficult to understand how they could have been shaped by the normal forces of erosion. When they came to examine their gully-riven fronts in detail, they discovered that the rocks were full of the remains of coral. Once sea had covered this area and these cliffs were then reefs bordering deep fish-filled lagoons. Coral cannot grow in muddy water so the rivers, murky with sediment from the land beyond, maintained the gaps in the reefs. Slowly the lagoons filled with accumulated sediments and the sea retreated. Eventually the level of the whole Australian continent rose. Rain and rivers eroded the soft sandstone that had filled the lagoon basin, scouring it away so that today the reefs face, not the sea, but desert covered with clumps of spinifex grass and stunted mulga trees. At their foot, on what was once the seafloor, lie particularly hard rocky nodules. From the ends of some of them project groups of thin blade-like bones. The geologists took the nodules back to their laboratory and soaked them for months in acetic acid. Gradually the rock fell away and exposed, in astonishing perfection, the first complete and undistorted skeletons of the world's earliest true fish.

Many more fossilised fish have been found dating from this very early period, not only in Australia but all over the world. Modern scanning techniques now make it possible to peer inside nodules and examine the fossilised remains they contain without dissolving away the stony matrix. Many different species have now been identified. Most, like their predecessors, were armoured in some way, with heavy scales attached to bony plates in the skin and fearsome teeth in their jaws. Some also had the beginnings of an internal bony skeleton including the elements of a vertebral column running longitudinally through the body and surrounding the primitive flexible rod. All of them had well-developed lateral fins, usually in two pairs – the pectorals just behind the throat, and the pelvics near the anus. There were, however, many variations. One form had a whole line of lateral fins. The pectoral fins of another were encased in tube-like bones and looked like probes or props. Some were bottom-living, some free-swimming, and one or two were gigantic, reaching 6 or 7 metres in length. In the face of this competition, nearly all the jawless proto-fish died out.

About 450 million years ago, a split appeared in the fish dynasty. An examination of living species reveals what happened. A set of genes in one group of fish for some reason became duplicated and this resulted in them producing bone in their skeletons. Their descendants became the ancestors of all backboned animals alive today – including ourselves. The other group used a softer, lighter and more elastic material to support their skeletons – cartilage. The descendants of this group are the sharks and rays. This ancient split in the ancestry of fish means that you and I are more closely related to a cod than the cod is to a shark.

The reduction of bone in the bodies of the early sharks doubtless made them considerably lighter, size for size, than their ancestors. Even so, muscle and cartilage is heavier than water, and to remain above the seafloor, sharks have to keep swimming. They drive themselves through the water in the same way as their ancestors, by the sinuous motion of the rear half of their bodies and the powerful thrash of their tails. But with the thrust coming from the back, the body is nose-heavy and liable to dive downwards. To correct this, a shark has two pectoral fins spread horizontally like the vanes of a submarine or the wings of a rear-engined aircraft. These fins are, however, relatively inflexible. The shark cannot suddenly twist them to a vertical position to act as brakes. Indeed, a charging shark cannot stop, it can only swerve away to one side. Nor can it swim in reverse. Furthermore, if it stops beating its tail it sinks. Some species, indeed, take rests at night and slumber, settled on the seafloor.

One branch of the cartilaginous fish has adopted this position more or less permanently, abandoning the energy-consuming labour of perpetually beating their tail to maintain themselves in mid-water. These are the rays and skates. Their bodies have become greatly flattened, their pectoral fins enlarged into undulating lateral triangles that have taken over the function of locomotion. The tail, therefore, need no longer beat. It has lost nearly all its muscles and become thin and whip-like, sometimes with a venomous spine at the end. The method works very well but it cannot produce the speed possessed by the free-swimming sharks. But the rays do not need that. They are not active hunters and live largely on molluscs and crustaceans. These they grub up from the seafloor and crush in their mouths which open on the underside. This position of the mouth is convenient for feeding in this way, but it complicates breathing. Sharks take in water through the mouth, pass it over the gills and expel it through slits. Were the rays to take in water in the same way, it would be full of mud and sand. So instead they have two openings on the upper surface of the head that channel water straight to the gills. It is then expelled on the underside through the gill slits.

One kind of ray, the manta, has reverted to swimming in the surface waters. The lateral extensions of its body enable it to remain aloft with only a small expenditure of energy, using the water for support in the same way as gliders use air. But undulating side-wings are not such powerful propellants as a thrashing tail, so the manta cannot swim as fast as its shark cousins or rival them as hunters. Instead, it sails slowly through the water on flapping wings, sometimes as much as 7 metres across, its immense slot-like mouth held wide open, gathering by filtration the floating shoals of tiny crustaceans and small fish.

The descendants of the great group of fish that strengthened their skeletons with bone today dominate the waters of the world. They solved the problem of their weight in a roundabout but very effective way. Several families spread from the open seas into coastal waters and eventually into shallow lagoons and swamps. Breathing for a fish is difficult in such places. The warmer water becomes, the less oxygen it can hold in solution. The open seas never become hot, but shallow waters do and in consequence become oxygen-poor. So when fish came to live there, they had to develop additional ways of getting oxygen. The bichir, an odd-looking elongated fish with up to eighteen dorsal finlets down its back, still demonstrates the method they used. It lives in the rivers and swamps of Africa and regularly rises to the surface of the water to take a gulp of air. This goes down its throat and into a pouch that opens from the top wall of its gut. The walls of this pouch are thick with blood capillaries which absorb gaseous oxygen. The bichir, in fact, not only has gills like any other fish, but a lung as well.

But an air-filled pouch brings other incidental advantages. It provides buoyancy, and this, for the bulk of the descendants of these air-breathing pioneers, became a more important faculty. With a bag of air inside them, they could float in the water without perpetually labouring their tails. Soon species appeared which could fill their air-bags by diffusing gas into them from the blood rather than by rising to the surface and swallowing air. In some cases the tube connecting the bag to the gut became no more than a solid thread. So the fish acquired a swim bladder and many different species with such structures swam in the lagoons beside the Gogo reefs.

The techniques of swimming were now revolutionised. By diffusing gas in or out of the bladder or expelling it directly through the connecting tube, a fish could accurately control its level in the water. Its pectoral fins, freed from the job of providing lift, could be used to refine the fishes' control of movement and their swimming skills reached near-perfection.

Water is 800 times as dense as air, and the slightest bump or protuberance on the body can cause drag, more even than it would do on a bird or an aeroplane. So, under the relentless pressure of natural selection, ocean-going fish that use high speed to chase and catch their prey – tuna, bonito, marlin, mackerel – have evolved the most marvellously streamlined bodies, sharply pointed in the front, swelling quickly to maximum diameter and then tapering elegantly to the two-bladed symmetrical tail fin which is their propellor. The whole of the rear half of the fish is in effect the engine for this propeller. Banks of muscles are attached to the backbone so that the tail can be beaten from side to side with unflagging strength throughout the fish's life. The scales, so heavy and rough in the early forms, have now become thin and smoothly fitting or been lost altogether. The surface is lubricated by mucus. The plate covering the gills fits closely to the body and the eyes barely bulge above the smooth contours. The pectoral and pelvic fins and the dorsal along the crest of the back play no part in propulsion. They serve only as rudders, stabilisers or brakes. When the fish is moving at speed and they are not required, they are clamped to the fish's side, fitting exactly into depressions and grooves on the surface. And along the top and bottom edge of the body, on either side of the tail, are tiny triangular blades that serve as spoilers to prevent turbulence.

The perfection of this shape is attested by the fact that species belonging to quite different families of fish have acquired it and thus bear a strong resemblance to one another. Once a species moves into the open ocean, and relies on great speed, either to eat or to avoid being eaten, the ruthless selections of evolution refine the fish's shape towards this, the most efficient, the most mathematically perfect form for the purpose.

Some species of surface-living fish, in danger of being outpaced by the hunters, have turned their pectoral fins to a special purpose. When pursued, they shoot out of the water and spread greatly elongated, broad pectorals which until now have been held close to the body. As the air catches the membranes, the fish is lifted above the waves and it glides for hundreds of metres, leaving its pursuers baffled. Sometimes as they fly, they tilt their bodies so that their tails dip into the water, and beat a further few strokes, renewing their impetus and extending their flight.

Not all fish have adopted a life of speed. Those living in mid-water or along the shores have different problems and requirements, but for them too, the acquisition of a swim bladder has had a potent effect on structure, for it has freed their body fins for all kinds of other purposes. Those of a pike have become elegant filmy sculls, rotating slowly back and forth from a joint within the body, so that the fish can compensate for the tiniest variation of current and hang above a rock as though it were suspended from an invisible wire. Gouramis have turned their pelvic fins into long thread-like feelers with which they explore the water ahead of them and, at breeding times, caress their mates. The dragon fish has expanded them into spectacular defensive weapons, each ray barbed with venom.

Several species, since body weight is no longer such a problem, have once again taken to armour. The box-fish, in the highly populated and potentially dangerous world of the reef, sails over the coral packaged in a crate of bone, its pectorals whirling, its tail fin flickering. The seahorse is also armoured and stiff-bodied. Its tail has no fin on it but is used as a mobile hook with which the fish anchors itself to weed or coral. Its body is held upright and what was the dorsal fin has become an undulating rear engine, which, with the help of whirling pectorals on either side, enables it to move erect and stately through the forest of corals and weed. The trigger fish eats coral, crunching the stony branches and extracting the little edible polyps. It has concentrated its finnage on its rear half, with a large flapping dorsal next to the tail and an equivalent one on the underside. This keeps its head free so that it can thrust it deep between the branches of the coral and select a particularly toothsome piece. The trigger, which gives the fish its name, is the leading ray of its dorsal fin, which has become bony. The two rays behind it have been turned into a locking device on the joint of its base. When waves crash over the reef and free-swimming creatures can be hurled against rock or coral, the trigger fish swims into a crevice, sticks up its bony trigger and locks itself in place so firmly that neither ocean currents, hungry predators nor inquisitive skin divers can extract it.

Some bony fish have emulated the cartilaginous skates and rays and taken to the bottom-living life, abandoning the swim bladder that, in their past, had been the source of their success. Their pectoral fins have been turned to yet more purposes. The gurnard has dispensed with the membrane on the front part so that the rays are free and can be moved independently rather like the legs of a spider. It uses them to turn over stones when it is looking for food. The flounder has become adapted to bottom-living to a quite extraordinary degree. It illustrates again the tendency of some creatures to show aspects of their evolutionary past during their own development, for when it hatches, it swims above the seafloor just as its ancestors undoubtedly did. After a few months, however, it undergoes a transformation. It loses the swim bladder it has had until now. Its head becomes twisted and the mouth moves sideways. One eye shifts right round the body so that it takes up a position alongside the other. Then the fish descends to the bottom and lies on its side. The pectoral fins are now of little use, though the fish still retains them. It swims by undulating the much enlarged dorsal and anal fins that fringe its sides.

So, driven by their beating tails, sculled by their pectorals, planing on lateral fringes, the fish swim with speed and accuracy through all the varied habitats of the sea, from the rococo constructions of the reef to the mountains and plains of the seabed, from the swaying forests of kelp to the blue sunlit waters of the open ocean. But mobility demands sensitivity: if you travel, one way or another you must be aware of where you are going.

All fish have one sense for which we have no parallel. Down their flanks and branching over the head, runs a line with a slightly different texture from the rest of the body. It consists of a number of pores, connected by a canal running just below the surface. This lateral line system enables the fish to detect differences of pressure in the water. As it swims, a fish creates a pressure wave that travels ahead of it. When this meets some other object, the fish by means of its lateral line can detect the change. This ability to feel at a distance also enables it to detect the movements of other fish swimming alongside it, an important ability for those species that form shoals.

The fish's sense of smell is acute. The nostrils open into cups that can detect the most minute changes in the chemical composition of water. Sharks can detect the taste of one part in 25 million so that, when the current is in their favour, they can smell blood from a body nearly half a kilometre away. They rely greatly on smell to guide them to food, and this might provide an explanation for the shape of that most grotesque of sharks, the hammerhead. Its nostrils are placed at the ends of two extremities that grow out from the side of its head. If it scents its prey, it swings its head from side to side to determine the direction from which the smell is coming. When it is equally strong in both nostrils, then the hammerhead swims straight ahead – and is often one of the first predators to reach the scene.

Fish are likely to have been able to detect sound from a very early period. The capsule containing the two arching semicircular canals that are found in either side of the skull of the proto-fish and the lamprey has been improved considerably by the jawed fish. They have a third canal in a horizontal plane and beneath it a large sac. All three canals and the sac have very sensitive linings and contain small limy particles which move and vibrate. Sound travels better in water than in air, and as the fish's body contains a high proportion of water within it, the sound waves penetrate the skull and reach the semicircular canals without the aid of the special passage needed by vertebrates that live in air. So fish are aware of the popply, slapping noises that other fish make as they travel at speed through the water, of the clicks made by crustaceans snapping their hard shells and the scrapings made by fish grazing over the coral.

The acquisition of a swim bladder brought the possibility of further improvements in both receiving and transmitting sound. Several thousand species of fish have developed bony connections of some kind to link their swim bladders to the inner ear capsules, so that the vibrations picked up and amplified by the sympathetic resonance of the swim bladder are transmitted to the semicircular canals. Some fish have also developed special muscles so that they can vibrate the swim bladder and produce a loud drumming noise. Catfish of several species do so and appear to be calling to one another as they move in murky water. Many other fish use sound to communicate with each other during courtship

Sight was also an ability acquired very early. The lancelet's eye-spot makes it aware of the difference between light and dark. The jawless fish, even though their heads were heavily plated, had chinks in their armour to accommodate eyes. Since the laws that govern the behaviour of light are universal, it is not surprising that the number of basic designs for an efficient eye are small. The trilobites produced a mosaic eye, as have insects. Otherwise, image-producing eyes, no matter which organism develops them, have a similar fundamental structure – a closed chamber with a transparent window and a lens in front and a photosensitive lining at the back. This is the pattern of the squid and octopus eye as well as of the mechanical one built by man, the camera. It is also the basis of the eye developed by the fish and bequeathed by them to all land-living vertebrates. The lining may contain two kinds of differently shaped cells, the rods and the cones. The first distinguish between light and dark, the second are sensitive to colour.

The eyes of almost all sharks and rays lack cone cells, so they cannot perceive colour. Not surprisingly, therefore, they are themselves drab creatures dressed in browns and greys, olive green and steel blue. When they are patterned, their designs tend to be simple spots and dapples. Bony fish, on the other hand, are strikingly different. Their eyes have both rods and cones, their colour vision is, for the most part, excellent, and their body colours are accordingly vivid and various. Sulphur-yellow fins are attached to sapphire bodies, orange spots are scattered on a sage-green flank, chocolate-brown scales are individually rimmed with peacock blue, tails are patterned like archery targets with a golden centre surrounded by outers of scarlet, black and white. There seems to be no pattern, no shade in the spectrum, that the bony fish have not deployed in order to decorate their bodies.

The most brilliantly patterned of all are those fish that live in clear sunlit waters where their designs are easily seen – in tropical lakes and rivers and, particularly and most lavishly, around coral reefs. Here, because of the abundance of all forms of life and the richness of food, there is a huge and crowded population. In such circumstances, species identification becomes very important and the fish have adopted the most vivid liveries to assist in it.

One group of fish, called, because of the beauty of their coloration, butterfly fish, show how diverse such patterning can be within one small family. They are all about the same size – only a few centimetres long – with roughly the same shape, slim, approximately rectangular, with high foreheads and small pouting mouths. Each species has its own particular place on the reef, with its own favoured depth and preferred source of food. One has elongated jaws for picking between the coral stems, another may specialise in cropping a particular kind of small crustacean. It is in the interests of each individual therefore to proclaim clearly among the confusion of swarming fish that its particular niche is occupied so that no other individual of the same species will poach its territory. Equally, the colours will draw the attention of a female to the presence of a male of the only kind with which she can have a fertile union. In many environments, the need to advertise in this way is limited by the danger of becoming a conspicuous target for a predator. For the butterfly fish, this risk is small, for, hovering over the coral, it can dart to safety among the stony fronds within a fraction of a second. So each species of the family, on the near-identical canvas of its body, carries a vivid and individual design based on stripes and patches, dots, eye-spots and zigzags.

As spawning time approaches, the need to identify a mate becomes particularly intense. Away from the reefs, in more dangerous and exposed waters, the males still often adopt brilliant colours, risking conspicuousness, in order to threaten their rivals and attract females. Pigment granules diffuse within their skins as they become excited and they fight with their colours, circling one another and flexing and quivering their fins like bullfighters' capes. They beat their tails and send pressure waves along the lateral lines of their rivals. They tear at the patterns on one another's fins. Eventually, when one has had enough, he signals submission by contracting the pigment in one set of cells and expanding that in another so that his flank patterns change and he hoists the flag of surrender. The winner is now free to court his female. He then uses much the same repertory of colours and patterns and fin displays as he did for aggression, but in a female these trigger a series of different responses that eventually culminates in the laying of eggs.

The eyes of some fish enable them to see not only what is going on in the water around them but in the air above the surface. The archer fish is partial to flies and other insects that may settle on plants growing on the banks. It takes aim, compensating for the way that light bends as it passes from water to air, and squirts a jet of drops, knocking the insect from its foothold so that it falls into the water and can be eaten. A small fish from Central America is even more specialised. It has a horizontal division across its pupil which effectively gives it four eyes – the two lower

Above: Dickinsonia, *a 565 million-year-old fossil from southern Australia. In 2018, a molecule was identified in such a specimen suggesting that this organism was not a plant but an animal and the first yet positively identified as such. (© Ilya Bobrovskiy)*

Marine iguana, grazing seaweed underwater, Galapagos Islands. (Pete Oxford/naturepl.com)

Long-necked giant tortoise, from Espanola, one of the more arid of the Galapagos Islands.
(Tui De Roy/Minden Pictures/FLPA RM)

Opposite top: *Near-horizontal layers of rocks of the Grand Canyon cut through by the Colorado River, Arizona, USA. (© Dean Fikar/Getty Images)* Opposite below: *A diver inspects a giant barrel sponge, Philippines. (Jurgen Freund/naturepl.com)* Above: *Hot spring with water coloured by bacteria, Yellowstone National Park, Wyoming, USA. (Floris van Breugel/naturepl.com)*

Opposite: *Portuguese man o'war, a colonial jellyfish. (Jurgen Freund/naturepl.com)*
Above: *Purple sea pen, Indonesia. (Georgette Douwma/naturepl.com)*

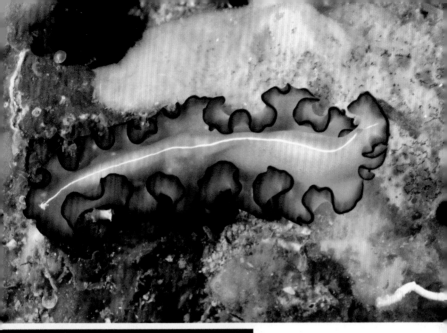

Opposite: *Lyretail coralfish over coral reef with soft corals, Red Sea, Egypt. (Georgette Douwma/ naturepl.com)*

Above: *Flatworm, Indonesia. (Franco Banfi/naturepl.com)*

Left: *Living crinoid, 180 metres down in the Caribbean Sea. (Doug Perrine/naturepl.com)*

Above: *Nudibranch in a Scottish loch. (Alex Mustard/naturepl.com)* Below: *Nautilus on a coral reef at night, Pacific Ocean. (Jurgen Freund/naturepl.com)* Opposite: *Bigfin squid at night, Indonesia. (Alex Mustard/naturepl.com)*

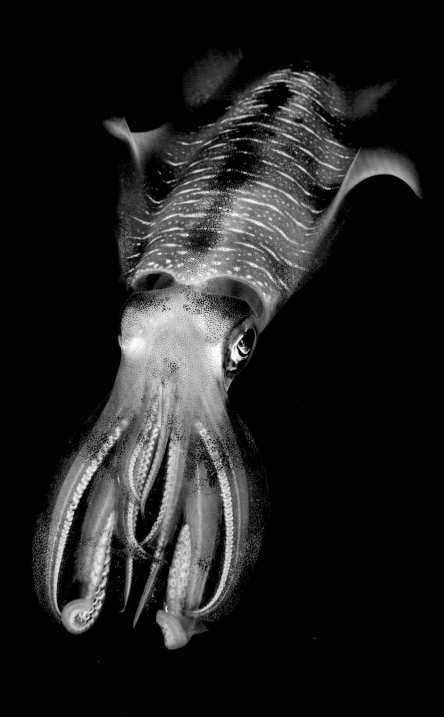

Above: *Group of cushion starfish on the sea floor, Galapagos Islands. (Brandon Cole/naturepl.com)*

Right: *The tube feet of a red cushion starfish, Florida, USA. (Andrew J. Martinez/ Science Photo Library)*

Above: *Crinoid sitting on a gorgonian with a soft coral in the background, Andaman Sea, Thailand. (Georgette Douwma/Science Photo Library)*

Left: *Velvet worm, New Zealand. (Alex Hyde/naturepl.com)*

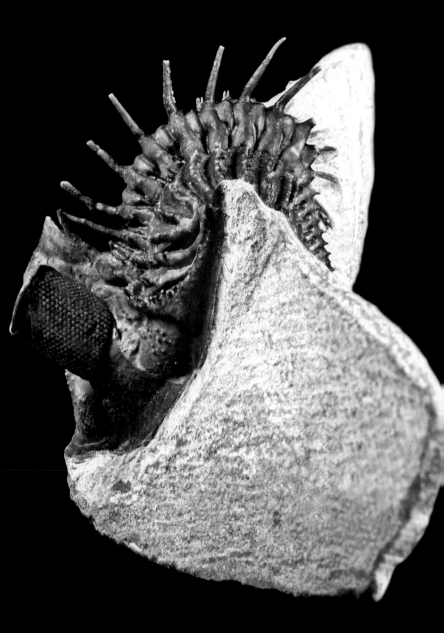

Opposite: *Tower-eyed trilobite, from 400 million-year-old sandstone, Morocco. (Natural History Museum, London/Science Photo Library)*

Left: *Horseshoe crabs spawning in the sand at high tide, Cape May, New Jersey, USA. (Sean Crane/Minden Pictures/FLPA RM)*

Left: *Robber crab descending a coconut tree, Aldabra, Indian Ocean. (Pete Oxford/naturepl.com)*

Above: *Cactus growing on a lava field, Galapagos Islands. (Kerstin Hinze/ naturepl.com)*

Right: *Spore capsules of apple moss, Inverness-shire, Scotland. (Duncan McEwan/naturepl.com)*

Left: *Scorpion stinging a spider, southern Europe. (Stephen Dalton/naturepl.com)*

Below: *Male wolf spider (right) courting a female by waving his palps. (Alex Hyde/naturepl.com)*

Horsetails, Oregon, USA. (Visuals Unlimited/naturepl.com)

Hawker dragonfly perched on water dropwort, Ireland. (Robert Thompson/naturepl.com)

Right: *Ancient bristlecone pine tree, Nevada, USA. (Kirkendall-Spring/ naturepl.com)*

Below: *Longhorn beetle in flight, with wing-covers lifted to free the wings beneath. (Stephen Dalton/ naturepl.com)*

Above: *Giant flower of a Rafflesia, Sarawak, Malaysia. (Jouan Rius/ naturepl.com)*

Left: *Bee orchid in flower, Dorset, UK. (Colin Varndell/ naturepl.com)*

Above: *Ground beetle, France. (Pascal Pittorino/naturepl.com)*

Right: *Two-tailed pasha butterfly expanding its wings after emerging from its chrysalis, the empty shell of which hangs above, Italy. (Paul Harcourt Davies/naturepl.com)*

Above: *Comet moth
from the rainforest of
eastern Madagascar.
(Imagebroker/Alexandra
Laube/Imagebroker/
FLPA)*

Left: *Caterpillar of
the Chinese silk moth,
feeding on oak leaves.
(Visuals Unlimited/
naturepl.com)*

Above: *Nest of a termite colony with a high cooling chimney, Kenya. (John Downer/naturepl.com)*

Right: *Green tree ants pulling leaves together so that they can be fastened with silk and so form a nest, Northern Territory, Australia. (Ingo Arndt/naturepl.com)*

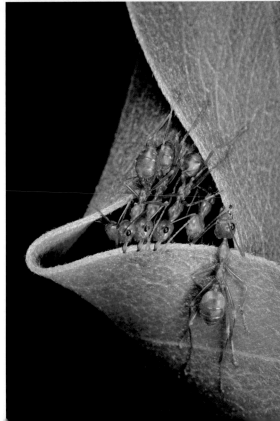

Left: *Sea squirt colony, Channel Islands, UK. (Sue Daly/naturepl.com)*

Below: *Larval sea squirt, stained to show its notochord. (Visuals Unlimited/naturepl.com)*

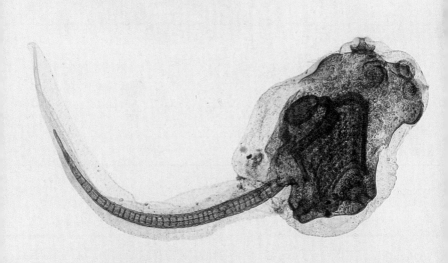

Manta rays chain-feeding on plankton, Maldives, Indian Ocean. (Doug Perrine/naturepl.com)

Sailfish attacking a bait-ball of sardines and sardinella off the Yucatan Peninsula, Mexico.
(Doug Perrine/naturepl.com)

Above: *Hammerhead shark with two small escort fish in shallow water, Bahamas. (Alex Mustard/naturepl.com)*

Right: *Seahorse, eastern Australia. (Alex Mustard/naturepl.com)*

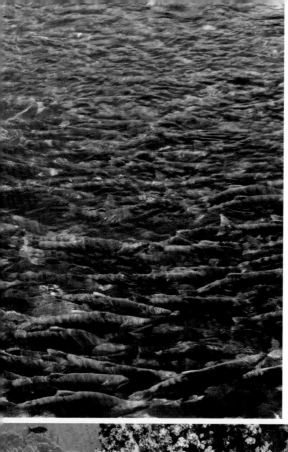

Left: *Sockeye salmon migrating up-river to spawn, Adams River, British Columbia, Canada. (Michel Roggo/naturepl.com)*

Below: *Adult emperor angelfish swimming past soft corals, Red Sea, Egypt. (Georgette Douwma/naturepl.com)*

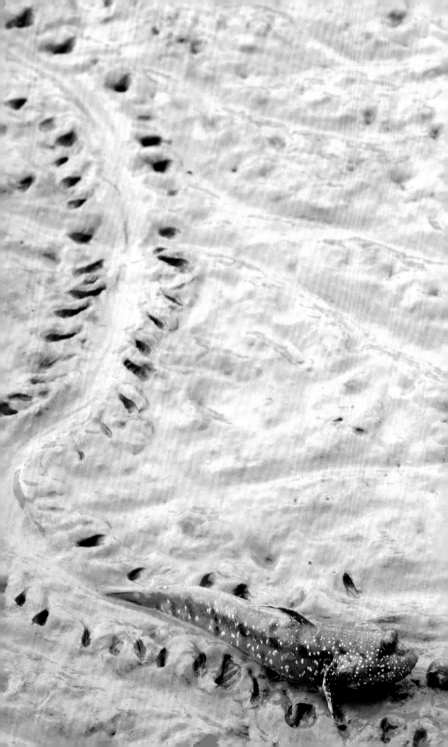

Opposite: *Mudskipper, using its pectoral fins to crawl across mud at low tide, Selangor, Malaysia. (Fletcher & Baylis/Science Photo Library)* Top: *Fossil lungfish, 385 to 359 million years old, Devonian rocks, Quebec, Canada. (Ken Lucas, Visuals Unlimited /Science Photo Library)* Above: *Albino axolotl. (Jane Burton/naturepl.com)*

Above: *Caecilian amongst rainforest leaf litter, Ecuador. (Pete Oxford/naturepl.com)*

Below: *Wallace's flying frog in mid-air. (Stephen Dalton/naturepl.com)*

knife fish in South America, elephant fish in West Africa, so called because they have an elongated lip like a small probing trunk. If you want to discover them, all you need is two wires at the end of a pole leading to an amplifier, powered by a small battery and attached to a little loudspeaker. If you dip the end of the wires in a stream where such fish are searching for food in the muddy bottom, you will hear a series of clicks. These are the electric signals, translated into sound so that they are detectable by human ears.

The fish have modified muscles in their flanks which generate and transmit these electric discharges. Some species send out signals almost continuously, others emit short bursts. Each seems to have its own identifiable code. The transmission creates flow patterns of current in the surrounding water. Any object with a conductivity different from that of the water will distort the pattern. The fish becomes aware of the change through receptor pores spaced out over its body, and even in the darkest, most Stygian waters, knows the shape and disposition of the objects around it.

The largest of such fish is the South American electric eel. It is not related to the true eels but looks superficially like them and so has acquired their popular name. It grows to a metre and a half in length and as thick as a man's arm. Often it makes its home in holes beneath a river bank or among rocks. Reversing into these holes, for a lengthy creature like an eel, must pose considerable steering problems. The eel does it with the aid of electricity. As you watch one tackling such a problem in a tank, you can detect the clicks of its discharges increasing in frequency as it identifies the outlines of the selected parking place behind it and slowly manoeuvres its great length into it without once touching the sides. But the electric eel has another set of batteries that produce not steady low-voltage direction-finding transmissions but sudden massive shocks so strong that if you pick up such a fish without the insulation of rubber gloves and boots, it can throw you flat on your back. The eel uses this kind of discharge for hunting. One reckless scientist recently used video analysis to discover exactly how the eel attacks its prey – using himself as its target. The eel leaps out of the water and then appears to nuzzle its way to its victim, releasing pulses of shocks as it goes. The scientist described these shocks as intensely painful, but he had sensibly chosen a small juvenile to study. A full-grown eel is one of the very few creatures in the world that can kill by electrocution.

Today, 500 million years after those jawless armour-laden creatures began to wag their tails and blunder over the muddy bottoms of the ancient seas, the fish have evolved into some 30,000 different species. Between them, they have colonised every part of the seas, lakes and rivers of the world. Their mastery of the waters is epitomised by that most splendid, valiant and efficient of fish, the salmon.

Five species of them visit North American rivers. They spend the bulk of their lives in the Pacific Ocean. When they are small, they feed on plankton. As they grow larger, they take to eating fish. In August each year, the fish that have just reached adulthood travel towards the American coast. They assemble off-shore and then begin to battle their way up the rivers, fighting and dodging the swift downward current, selecting with the help of the pressure-sensitive pores of their lateral lines the reaches where the current is marginally slacker, resting in quiet pools, recovering their strength before tackling another stretch of rapids.

These rivers are not chosen at random. Each salmon remembers the precise taste and smell of the waters in which it hatched, a sensation derived from the mix of minerals in its mud and the plants and animals that live in it. They can detect this flavour even when their home water is diluted out at sea to one part in several million. But to find a particular estuary across several hundred kilometres of ocean, they need some kind of map. It seems that the large-scale markers in the map they use are not physical or even chemical, but magnetic. They use the changing strength of the earth's magnetic field in different parts of the planet's crust to guide them back to a particular bay and then use their sense of smell. As the scent gets stronger and stronger, they locate one special river and swim up it until they find one particular stream. We know that it is smell that guides them at this stage, for salmon with blocked nostrils get lost. The accuracy of their memory and navigation is astounding. In multiple studies, many thousands of young fish have been marked soon after they hatched – only one or two return to a river other than the one in which they first swam.

The compulsion to return may be strong, but the obstacles are huge. The move from saltwater to fresh in itself requires major adjustments to the chemistry of the body, but the salmon manages to make them. On their way upstream, they may encounter waterfalls. Their sharp eyes select the lowest part of the lip of the fall. Then, flexing their powerfully muscled silver bodies, they thrash their tails and leap from the water. They may have to jump again and again before at last they land in the pools at the head of the falls and are able to continue on their journey.

Eventually they reach the shallow stretches where their parents spawned and there they lie resting, their heads pointing upstream, flank to flank and so thick that the pale sand of the river bed is hidden by the black of their backs. Now, within a few days, the shape of the males' bodies change with astonishing speed. They develop high humps on their backs. Their upper jaws become hooked and their teeth grow into long fangs. These are useless for feeding – but the time for feeding has long since passed. These teeth are for battle. The males wrestle and fight, flank to flank, seizing one another's jaws, striking at their opponents with their splayed teeth. The water is so shallow that their writhing humped backs break clear of the surface. At length, one wins and claims a scrape in the gravel. A female joins him. Swiftly eggs and sperm are shed and sink beneath the gravel grains.

Now the adults are totally spent. They do not have enough energy even to heal their battered wounded bodies. Their scales fall off, the once powerful muscles dwindle and they die. Not a single one of the millions of fish that fought their way up the river ever returns to the sea. Their frayed bodies lie rotting in drifts on the surface of the streams and are washed up in piles on the sandbanks. Here and there, a last survivor makes a few final despairing flaps. Gulls gather in flocks to peck out eyes and strip the yellowing flesh.

But in the gravel, the eggs remain, a thousand or so from every female. They stay safe throughout the hard winter. Next spring they hatch. The fry remain in the streams for a few weeks, feeding on the flush of insects and crustaceans that appear in the warming waters. When they are finger-lings, they leave, following the current downstream to the sea. Some species will swim there for two seasons, others for up to five. Many will be eaten by other fish in the sea, but eventually the survivors will fight their way back up their own river to spawn and then die in the very place where they were hatched.

Three-quarters of the world's surface is covered by water. Three-quarters of the world belongs to the fish.

SIX

The Invasion of the Land

One of the most crucial episodes in the history of life took place some 375 million years ago in a tropical freshwater swamp. Fish began to haul themselves out of water and become the first backboned creatures to colonise the land. To cross this frontier, they, like, the first terrestrial invertebrates, had to solve two problems: first, how to move around out of water, and second, how to obtain oxygen from the air.

There is one fish, alive today, which manages to do both these things – the mudskipper. It is not closely related to those fish that pioneered the land, so any comparisons with them have to be made with caution, but even so it can give us a hint about how that momentous move was accomplished. Mudskippers are only a few centimetres long and you can find them in mangrove swamps and muddy estuaries in many parts of the tropics, lying on the glistening mud well beyond the lap of the waters. Some may even be clinging to the arching aerial roots of the mangroves or clambering up the trunks. A sudden movement or an abrupt noise will send them skittering back to the safety of the water. They come out to feed on the insects and other invertebrates that swarm on the soft oozy surface of the mud. They move by suddenly flexing the hinder end of their body so that they give a little skipping jump. But they also have a steadier, more sober way of edging themselves forward with their front pair of fins. Each of these has a fleshy base supported internally by bones – the fin is, in effect, a rigid crutch, and with it the fish can lever itself forward.

But what was the backboned creature that first made this momentous move? Fins with muscled fleshy bases, evolved so recently in evolutionary history by mudskippers, were also possessed by a varied group of fossil fish that flourished from 450 to 70 million years ago. One of the commonest was a family called the coelacanths. So when, in 1938, a living coelacanth was hauled up off the coast of South Africa there was great excitement among evolutionary scientists. Here was a chance to discover the details of the flesh and internal anatomy of these ancient fish that no fossils could provide. Unfortunately, however, the fisherman had already gutted the fish and thrown away the all-important entrails. A search was mounted all along the coast of southern and eastern Africa for another specimen. Substantial rewards were offered. But none appeared. Then in 1952 a second coelacanth was discovered. This time it was caught off the coast of the Comoro Islands, north of Madagascar. The local people apparently did not value these strange creatures particularly highly. Their flesh was not unusually delicious and, once hooked, they fought hard. But even so, one or two were landed every year and soon scientists had numerous specimens to examine in detail. Then another population of coelacanth was discovered in Indonesia. Underwater cameramen in submersible craft filmed them, showing how they swam. They moved in a slow and stately way, sometimes punting themselves across the seafloor with their strange fleshy limbs. But detailed anatomical studies, including genetic analysis, led scientists to conclude eventually that while the coelacanths are undoubtedly an extremely ancient group, they are not as closely related as had been thought to the first backboned animals that colonised the land.

So the search was renewed for the common ancestor of terrestrial vertebrates and in 2004 an even more likely candidate was discovered, this time by palaeontologists. A team working on Ellesmere Island in northeast Canada unearthed the fossilised bones of a large, heavily built fish-like creature. The local Inuit people, looking at the remains, called it Tiktaalik, the name they use for the burbot, one of the bigger of the local freshwater fish that they regularly catch, and this now became the creature's scientific name. But in fact, Tiktaalik is not like any fish – or indeed anything else – that is alive today.

It was around two metres long and had a massive flattened head not unlike a crocodile's with a pair of eyes on its upper part. Its body was protected with fish-like scales and it had the beginnings of a neck. But its importance in evolutionary history lies in its limbs. Those at the front were each fringed with a fin, as you might expect from a fish. But their fleshy base contained bones that were jointed both at the elbow and at the wrist. This clearly was a limb that could have helped its owner to move about out of water. But why should it have done so? Perhaps it was to collect food – either the remains of dead marine creatures that were washed up on to the beach, or perhaps to catch some invertebrates that by now were well established farther inland.

But how did it breathe out of water? The mudskipper manages to do so by holding water in its mouth which it swills over the lining of its mouth with a rolling action of its head to extract the oxygen. It also absorbs some directly from the air through its moist skin. But these devices only allow it to remain out of water for a short time. Within a few minutes it has to return to wet its skin and take a fresh mouthful of water. The surviving coelacanth certainly cannot suggest an answer, for today it never leaves its deep waters. Once again, however, there is a living creature that has a solution.

Many of the swamps around the flood plains of African rivers turn to hard sun-baked mud during the dry season, yet one fish, the lungfish, manages to live in them all year round and survive the dry season by breathing air. As the pools shrink, the lungfish burrows into the mud at the bottom.

There it curls into a ball, wrapping its tail around its head, and secretes mucus to line its hole. As the sun bakes out the last moisture from the mud, the mucus turns to parchment. Other primitive freshwater fish, such as the bichir, have a pouch opening from the gut which enables them to breathe air. The lungfish has a pair, and now, out of water, it is totally dependent on them. In burrowing down, the fish makes a tube through the mud a centimetre or two across. Air now passes down this to the mouth of the fish which is connected to tiny openings in the parchment cocoon. By spasmodically pumping its throat muscles, the fish draws air down its throat into its pouches. The walls of these are thick with blood vessels which absorb gaseous oxygen. These organs are simple lungs, and, with their aid, the lungfish can survive out of water for several months, even years.

When the rains do finally return and water fills the pond again, the fish, within the space of a few hours, comes to life, wriggles free of its cocoon and the re-softened mud and swims off. Now, in water, it breathes with its gills like any normal fish, but like the bichir it uses its lungs too, rising every now and then to gulp air from the surface, a talent which is particularly valuable when the water in the pools becomes tepid and loses most of its oxygen.

Four different species of lungfish are found in Africa, one in Australia and another in South America. They were, however, very much more abundant 350 million years ago and their fossils are often found in the same sort of deposits that contain coelacanths. The relationship between lungfish, coelacanths and land-living four-legged animals – tetrapods – has long been debated. Once again, molecular genetics provide the definitive answer. It so happens that lungfish have the largest known genome of any vertebrate – over ten times the size of a human's, and this makes it complicated to sequence, but studies of the lungfish's protein-producing genes have confirmed that it is more closely related to tetrapods than it is to the coelacanths. These studies also show that all three lineages diverged very rapidly around 380 million years ago, shortly after the time of Tiktaalik.

Scientists still do not know whether Tiktaalik was our ancestor or merely a distant cousin. Whatever the case, within a few million years of Tiktaalik waddling across the mudflats, tetrapods had become true land-dwellers. The swamps they inhabited were thick with horsetails and club mosses, both of which now grew to the size of trees. Their fallen trunks, accumulating in the swamps, eventually fossilised and turned to coal and so, not surprisingly, coal mines were the places where scientists first found the bones of those original vertebrate land-dwellers, the amphibians.

Some of them must have been terrifying. They grew to a length of three or four metres and their jaws were spiked with lines of cone-shaped teeth. They dominated the land for the next hundred million years. But then, around 200 million years ago, life on this planet suffered a global catastrophe comparable to the later more famous one that brought an end to the dinosaurs. About half of all the species on the planet disappeared. The last of the giant amphibians seems to have been a five-metre-long monster that lived about 110 million years ago in what is now Australia.

One kind of amphibian alive – the giant salamander – can give us some slight impression of those early forms. There are two species, one in China and the other in Japan, and both, unlike their ancestors, now spend their lives in water.They have a flat spade-like head, tiny button eyes, a long tail and a wrinkled warty skin that hangs in folds around the body. The Japanese species grows to a length of about a metre and a half, only a quarter of the size of its ancestors but exceptional for contemporary amphibians. Most salamanders and their cousins the newts are comparatively tiny. The biggest is only around 10 centimetres long. Collectively, they are known as the 'urodeles', the 'tailed ones'.

The legs of a newt, advanced though they are compared to the fins of a coelacanth or a mudskipper, are not very efficient. They are so short that to take a reasonable stride forward with its hindleg, a newt has to flex its body laterally. Most of its time is spent on land, hiding beneath stones or sheltering in damp mossy places searching for the worms, slugs and insects on which it lives. But it cannot stray far from water. For one thing, its skin is permeable and in consequence the animal, in a dry atmosphere, loses its body liquids very quickly and dies. To make matters worse, a newt, like other amphibians, lacks a mechanism for drinking with its mouth. It has to absorb all the liquid it requires through its skin. It must also keep its body moist to help it breathe. Its lungs are relatively simple and not totally sufficient for its needs, so its oxygen intake, like that of the mudskipper, has to be supplemented by absorption through its wet skin. Both these requirements restrict it, and most amphibians, to moist places. But there is a third need that ties it to water: its eggs, like those of a fish, do not have waterproof shells, so it somehow has to find water for them when the time comes to breed.

During its water-living phase in the breeding season, a newt becomes quite fish-like. It swims, with its legs held alongside its flanks and out of the way, by sinuous movements of its body and by beating its tail. The male of some species develops a crest along his back like a dorsal fin and becomes brightly coloured, as so many fish do during courtship. When he displays, he beats water with his tail and flexes his crest, sending powerful currents towards the female or rivals. These they detect with lines of sensors on the head and along the sides of the body which are an inheritance from the fish and an equivalent of their ancient lateral line system.

The female lays a great number of eggs, attaching each individually to the leaf of a water plant. When the young hatch, they are even more fish-like than their parents, for they have no legs and breathe not with lungs, which will develop later, but with feathery external gills. They are tadpoles.

Some salamanders in Central America exploit the possession of such a water-living larva to give themselves an alternative in the way to spend their adult lives. One species living in a lake in Mexico regularly changes to a land-living adult form in the normal fashion. But if there is a particularly wet season and its lake does not shrink and dry, then its larvae retain their feathery gills. They continue to grow well beyond the size at which they would normally change shape and become as big as, if not bigger than, the land-living form. Eventually, while still retaining their tadpole appearance, they become sexually mature and breed.

One closely related species has reverted permanently to the aquatic life of its ancestors. It always breeds in a larval condition, its external gills growing into great branching bushes on either side of its neck. It lives in Mexico, and in ancient times the local people, the Aztecs, perhaps recognising how odd this was, gave it a name which means 'water monster' – axolotl. Today it is almost extinct in the wild, but an albino form of it survives in considerable numbers in zoological laboratories. The fact that it is a salamander can be experimentally demonstrated by feeding it with thyroid extract. It will then lose its external gills, develop lungs and turn into a creature that closely resembles other salamanders such as the burrowing salamander that lives in Florida.

Farther north, in the United States, one amphibian has reverted irrevocably to water-living. This is the mud puppy. It has both gills and lungs, lays its eggs in a nest in the bottom of a stream and remains in water throughout its life. This species cannot be changed into a different form. Its body can detect the thyroid hormone, but for some reason the treatment does not affect the genes that control metamorphosis. The same is true of the olm, another blind salamander that lives in the underground rivers of Slovenia and which local traditions maintain are baby dragons.

Some salamanders have taken this reversion to a fish-like existence even further. They seem to be losing not only their lungs but their legs. The siren, a metre-long amphibian from the southern United States, has lost its back legs altogether, and its front legs are not only greatly reduced in size but have no bones within them, merely cartilage, so that they are of no practical use in locomotion. The amphiuma, from the same part of the world, still possesses all four of its limbs, but they are so minuscule that you have to look very carefully if you are not to miss them. Indeed, it is superficially so like a fish that it is known locally as the Congo eel.

This abandonment of both the major innovations made during the vertebrates' colonisation of the land occurs not only among those salamanders that have taken to water but even among some that spend their lives almost entirely on land. Many American salamanders have lost their lungs and yet manage to breathe adequately through their wet skin and the moist membranes lining their mouths. But this can only be done at the price of a restriction of body size. Breathing in this way will be more efficient if the body is of the size and shape that gives maximum skin area and minimum body volume, and this is indeed just what is found among these lungless salamanders; their bodies are thin and elongated and none of them grows to more than a few centimetres in length.

One group of amphibians, the caecilians, have lost their legs altogether and taken to burrowing underground in soft soil or mud. Their anatomy is so specialised and so different from the urodeles that they are classified in an order of their own. They live only in warm wet parts of the world, the bulk of them in the tropics. Not only do they lack legs but there is no sign of an internal girdle of bones at either shoulder or hip. They also have an extremely elongated body. Urodeles usually have a dozen or so vertebrae in their spines; a caecilian, however, may have as many as 270. Eyes are of little use burrowing underground, and often they have become covered in skin or even bone. Compensating for this loss of sight, some species have developed small extendable tentacles at the angle of their jaws which serve as sensitive feelers.

Caecilians reproduce in a way that is unusual. They do not return to water. Nor do they practise internal fertilisation. Instead, male and female meet in a damp burrow. The male protrudes a tube-like extension of his genital opening and inserts it into the female's vent and so fertilises her eggs. In some species the female then lays a string of soft-shelled eggs and guards them carefully in her underground chamber. In others, the young hatch within her and emerge from her body alive. Their mother then feeds them in a peculiarly intimate fashion – they nibble the fat-rich outer layers of her skin.

Caecilians are rarely seen, for they seldom come to the surface except at night, and even if they are accidentally dug up, they may well be mistaken for giant brightly coloured earthworms. But unlike earthworms, which eat rotting vegetation, the caecilians seem to be carnivores, feeding on insects and other invertebrates. They have hunter's jaws and when they suddenly expose a huge gape, they can be quite alarming – if you think you are handling an ordinary, inoffensive worm.

There are nearly 200 species of caecilians known and about 500 species of urodeles. But by far the most numerous amphibians alive today belong to a third group, the anurans, the 'tail-less ones'. There are about 5,500 of them. In temperate parts of the world, it is popularly thought that there are two kinds of anurans – those with smooth moist skins called frogs, and those with dryer, more warty skins, the toads. The distinction, however, is literally no more than skin-deep. In the tropics, where the bulk of the anurans live, there is no clear distinction between the two kinds, so a species could be called frog or toad with equal accuracy.

Instead of lengthening their bodies, like the caecilians, this group have shortened it. Their vertebrae have become fused together, and far from losing their legs, they have developed them enormously and some have become prodigious leapers. The biggest anuran of all, the Goliath frog from West Africa, is able to jump three metres or so. Spectacular though this is, many smaller frogs can easily outdo it, if their jumps are judged in relation to their body size. A few tree-living species can travel fifteen metres or so through the air, about a hundred times their body length, by becoming gliders. Their toes have become greatly elongated and are connected with webs of skin so wide that each foot has in effect become a small parachute. When the frog leaps off the branch of a tree, it splays its toes so that, instead of falling, it planes gently downwards.

The frog's leap is not merely a way of getting from one point on the ground to another. It is also a very effective method of escaping from an enemy – so explosive and surprising that catching a frog can be a difficult business, whether you are a human or a hungry bird or reptile. And as anurans, with their soft vulnerable bodies, are much

From the beginning of their history, the amphibians were hunters, preying on the worms, insects and other invertebrates that had preceded them on to the land. The vast majority of them remain so today despite the appearance of bigger and more powerful hunters which have compelled them to be more circumspect in their behaviour. Some are still quite formidable. The horned toad of South America has a gape so big that it can with ease engulf nestling birds and young mice. But no amphibian can truthfully be described as nimble and for hunting they have to rely on something other than agility – their tongue.

The extendable tongue is an amphibian invention. No fish ever had one. It is attached not to the back of the mouth as ours is, but to the front. In consequence, the frogs and toads can stick it out much further than we can, simply by flicking it forward – a useful talent for a rather slow-moving hunter without a neck. Its end is both sticky and muscular so that a toad can use it first to grasp a worm or a slug and then to carry it back to the mouth. But most will only eat food that moves. Surround a toad with dead but palatable prey and it will just sit there – and starve.

Many amphibians, including the horned toad, have very serviceable rows of teeth on their jaws, as their ancestors had, but these are used for defence or as a way of gripping the prey. They do nothing to break up the food into easily swallowed gobbets or to tease out the hard inedible bits. No amphibian can chew. This is the reason why toads, when they seize one end of a worm, methodically rake the length of it with their forefeet to remove any bits of sticks or earth that might be stuck to it. The tongue helps

that detects low frequencies. Our own eardrums, despite the apparent similarities in structure, evolved completely separately from those of anurans which developed around 250 million years ago.

Frogs and toads today are most impressive singers. The lungs which blow air through their vocal cords are still simple and relatively feeble, but many frogs amplify the sound of their voices with huge swelling throats or resonating sacs bulging from the corner of the jaws. An assemblage of frogs, calling in a tropical swamp, can create such a noise that a human voice has to shout to make itself heard. The variety of sound produced by different species is enormous and amazing to anyone who has only heard frogs of the temperate regions. There are groans, metallic clicks, mewings and wails, belches and whinnies. It is intriguing to speculate, as you stand in a swamp listening to this astounding and deafening chorus, that, although much must have changed in the millions of years since the first amphibians appeared, it was, nonetheless, an amphibian voice that first sounded over the land which, until then, had heard nothing but the chirps and whirrs of insects.

The amphibian chorus, rising from a pool or a swamp, is a prelude to mating, a summons to all other members of the same species to assemble and breed. The great majority of amphibians still mate in water. Though males usually grasp the females, the act of fertilisation still, with very few exceptions, takes place outside the body. Sperm swim to the eggs, as fish sperm does, and for this process, water is normally essential. Once this has happened, the adults usually return to land.

Thus abandoned, the eggs are now beset by danger. Unprotected by a shell, they are easy meals for insect larvae and flatworms. Those that survive and hatch are then pounced on by water beetles, dragonfly larvae and many kinds of fish. The mortality is gigantic; but so is the number of eggs laid. A female toad may lay 20,000 eggs each season; perhaps a quarter of a million in her lifetime. Out of all these, only two have to live to maturity to maintain the level of the population. The strategy is an ancient one. Fish used it and still do. But it is an expensive one in terms of living tissue produced, and it is not the only one possible.

Some frogs have adopted a different technique. They lay comparatively few eggs but look after them carefully, protecting them from predators. The pipa toad is one of the most aquatic of the anurans, spending all its life in water. It is a grotesque creature with a flattened body and a squashed-looking head. When they mate, the male grasps the female with his arms, as most water-breeding anurans do. But then follows the most extraordinary and graceful ballet. The female kicks with her legs so that the pair soar upwards in an elegant slow somersault. As they descend, the female extrudes a few eggs which are immediately fertilised by the male's sperms that have been discharged into the water at the same time. Then, with delicate movements of his webbed hind feet, toes distended so that they form a fan, he gathers up the eggs and gently spreads them over the female's back. And there they stick. Again and again this arching leap is performed until a hundred or so eggs are fixed in an even carpet on the female's back. The skin beneath them begins to swell and soon the eggs are embedded in it. A membrane rapidly grows over them, and within thirty hours the eggs have disappeared from sight and the skin on the female's back is smooth and entire once more. Beneath the skin, the eggs develop. After a fortnight, the whole of the female's back is rippling with the movements of the tadpoles beneath. Then after 24 days, the young break holes in the skin and swim swiftly away to seek safe hiding places.

Other pond-dwelling anurans find safety for their brood in less extreme ways. Several simply find or manufacture private swimming pools. This is not so difficult in the tropical rainforest where the rainfall is so heavy and so well spread throughout the year that the centres of many plants are permanently filled with water. Members of the bromeliad family are shaped like great rosettes with deep water-filled centres. Some grow tall-stemmed on the ground. Others squat on the branches of forest trees with their roots dangling beneath them in the humid air. Their centres then become, in effect, miniature ponds high in a tree. No fish could possibly reach them. But frogs can, and several species in South America have taken up permanent residence. They lay their eggs in these chalices and there the young go through their entire development, sharing their pool with nothing more dangerous than a few innocuous insect larvae. In Brazil, another small frog builds its own pond on the margins of forest pools, constructing a crater ringed with a low mud wall about 10 centimetres high. The eggs are laid here and the tadpoles will stay in their privileged and exclusive bath until rain raises the level of the main pool and floods their quarters or breaks down the walls.

When the first amphibians appeared, there was, of course, one comparatively safe place for their eggs and young – the land. At that time no other vertebrates were there to steal eggs and gulp the larvae, and there were no risks comparable to those in the water posed by shoals of hungry fish or marauding voracious arthropods. If the amphibians could manage to deposit their eggs out of water, their young would certainly have greatly increased chances of survival. But there were problems. How could the eggs be prevented from drying out and how could tadpoles develop out of water? Whether any of those first amphibians overcame such difficulties, we do not know. Had they done so, their fossils would doubtless have been found far from the traces of lakes, swamps, streams and pools. Today the attraction of using the land for breeding is not so great, for the amphibians no longer have it to themselves. There are reptiles, birds and even mammals that relish amphibian eggs and tadpoles if they can find them. Nonetheless, even today, many frogs and toads find it advantageous to take the risk and follow this strategy.

One European species, the midwife toad, spends most of its life in holes, not far from water. It mates on land. The male calls to the female from the edge of his hole, making a repetitive high-pitched sound that is amplified by the subterranean echoes coming from his hole and resounds eerily in the late spring night. The female joins him, and, as she extrudes her eggs, the male fertilises them. A quarter of an hour later, he begins to take up the strings of eggs, twining them around his hindlegs. For the next few weeks, he hobbles about with them wherever he goes. If his surroundings get dangerously dry then he moves to moister ones. Eventually, when the eggs are about to hatch, he hops down to the edge of a pool and dips his legs with their burden of eggs into the water. He stays there for the hour or so that is necessary for all the tadpoles to emerge and then returns to his hole. The species may be called the midwife toad, but it is in fact the male that ensures the safe arrival of the young.

The South American poison frogs have a variation of the same technique. Their eggs are also laid on moist ground and the males crouch beside them on guard. When the tadpoles hatch, they immediately wriggle to the male and climb on his back. His skin secretes a great deal of mucus which both keeps the young attached and prevents them from drying out. They have no gills, but obtain their oxygen by absorbing it through the skin of their bodies and greatly enlarged tails.

In Africa, there are frogs that manage to breed on the branches of trees. They select one that overhangs water. They couple and the female begins to excrete a liquid from her vent which both she and her mate beat into a lather with their hindlegs. The eggs are then laid in the resulting ball of froth. In some species the outside of the suds hardens into a dry crust, retaining the moisture within; in others, the female regularly descends to the pond or stream beneath, absorbs water through her skin and then returns to the egg mass and moistens it with her urine. The eggs hatch and the young tadpoles develop within the froth until, at the appropriate time, the lower part liquefies and the tadpoles drop out and fall into the water below.

Other frogs avoid the need to provide water for their tadpoles by producing young that complete the whole of their development within the egg membrane. This, however, makes it impossible for them to feed during the larval stage, as free-swimming tadpoles can do, so they have to be provided with specially large quantities of yolk so that they can nourish themselves. This, in turn, means that the female can only lay a relatively small number of eggs in a clutch. The whistling frog from the Caribbean, which practises this technique, lays only a dozen or so, which it places on the ground. Development is very rapid and within twenty days each egg contains a tiny froglet which pierces the egg membrane with a minute spike on the tip of its snout and so emerges, having dispensed with external water altogether.

The most extreme and physically complicated breeding techniques are those in which the eggs and the developing larvae are kept moist by being retained actually within the body of the parent. The female of one South American frog called *Gastrotheca* has a brood pouch on her back with a slit-shaped entrance. When the pair begin to spawn, the male, which is smaller than she is, climbs on her back and clasps her around the throat. She then raises her hindlegs so that she is crouching with her nose down and her back tilted. One by one she extrudes eggs. The male fertilises them and they roll down a moist groove and into the brood sac. There they develop and hatch. One species of *Gastrotheca* produces about 200 young at a time. These emerge and are released into water as tadpoles. Another species, however, has only about twenty young but provides them with more yolk each and they remain within the sac until they become froglets. The female releases them by reaching forward with her hindleg, inserting her longest toe into the sac entrance and pulling it so that it enlarges and her young are able to clamber out.

The most bizarre of all these techniques, at least to our eyes, prejudiced as we are to a mammalian way of doing things, is that practised by Rhinoderma, a tiny frog that Darwin found in southern Chile. When the females have laid their eggs, which they deposit on the moist ground, the males sit in groups around them on guard. As soon as the developing eggs begin to move within their globes of jelly, the males lean forward and appear to eat them. Instead of swallowing them, however, the eggs are taken into the vocal sac which is unusually large and extends right down the underside of the male's body. There they develop until one day the male gulps once or twice, suddenly yawns, and a fully formed froglet leaps out of his mouth.

The acme of parental care among amphibians, however, is that provided by a West African species of Nectophrynoides, the females of which retain their young inside their bodies in a way that compares very closely with the technique of placental mammals. These toads are only about 2 centimetres long. Most of the year they are hidden in rock crevices. When the rains come, however, they emerge in great numbers and mate, the male clasping the female around the groin. Their vents are pressed together so that the sperm can make its way into the female. The fertilised eggs are not then laid but remain inside the female's oviduct. The tadpoles that develop from them are complete with mouths and external gills and they feed within the oviduct on tiny white flakes secreted from its walls, nibbling them just as though they were independent creatures browsing in a tiny pond. When, after nine months, the rains at last return, the female gives birth. Her stomach and oviduct do not have muscles that can contract and so expel her young as a mammal's womb has. Instead, she gives birth by bracing her body against the ground with her forelegs and then inflating her lungs so that they swell into her abdomen and squeeze the young out by pneumatic pressure.

So by these and many more ingenious techniques, the anurans have minimised their dependence on moisture for mating and for hatching and rearing young. Their permeable skins, however, still dictate that their surroundings are to be moist if the animals are to avoid death by desiccation. But one or two species have succeeded in minimising even this requirement.

There could scarcely be a less promising environment for an amphibian than the desert of central Australia where sometimes several years may pass without any rain falling. And yet a few kinds of frogs manage to live even here. The water-holding frog, Cyclorana, appears above ground only during the brief and infrequent rainstorms. Water then may lie on the rocks of the desert for several days, even a week or so. With frantic speed, the frogs feast on the great flush of insects that have also come with the rain. And they mate, laying their eggs in the shallow tepid pools. The eggs hatch and the tadpoles develop at a spectacular rate. As the rainwater soaks away and the desert once again dries, the frogs, adults as well as young, absorb water through their skins until they are tightly bloated and almost spherical. Then they burrow deep into the still soft sand and excavate a small chamber. Here they secrete a membrane from their skin so that they resemble a plastic-wrapped fruit from a supermarket. This effectively prevents water loss by evaporation through the skin, though the animal must doubtless lose some moisture by breathing, which it is able to do through tiny tubes attached to its nostrils and opening through the membrane. It can remain in this state of suspended animation for at least two years. The technique is very reminiscent of that used by the amphibians' far distant and antique cousin, the lungfish.

Nonetheless, the fact remains that even this frog is dependent upon rains arriving at some time, and its active life is, in reality, condensed to that brief moment when the desert is wet. To survive, remain active and breed in areas where there is little or no rain and no open water at all, a creature must have both a watertight skin and a watertight egg. It was the acquisition of these two characteristics that constituted the next great evolutionary breakthrough.

SEVEN

A Watertight Skin

If there is one place on earth where the reptiles still rule, it must be in the Galapagos Islands, isolated in the emptiness of the Pacific, 1,000 kilometres from the coast of South America. The reptiles reached them long before human beings and other mammals arrived four centuries ago. Those reptiles must have drifted there as involuntary passengers on the great rafts of vegetation that float down the rivers of South America and are swept out to sea. We have since introduced many other mammals, but even now there are small remote islands in the group where the rocks are still covered with herds of lizards, where giant tortoises lumber through the cactus and where you feel, as you land, that you have stepped back 200 million years to a time when such creatures dominated the planet.

The Galapagos lie across the equator, roasting in the sun. They are all volcanic. The larger ones rise almost 3,000 metres, so high that they attract clouds and produce their own rain; their flanks, as a result, are thinly covered with cactus and straggling, dusty bush. The smaller islands, however, are largely waterless. Their extinct craters are surrounded by congealed lava, its surface rippled by the corded swirls and bubbles that formed when it oozed like treacle out of the vents. On the few occasions that rain falls here, it runs off the rock and disappears almost immediately. There are no trees or bushes to give shade, only a few fingers of cactus, furred with spines. The black lava, grilling in the sun, is so hot that it is painful to touch with your bare hand. An amphibian here would be shrivelled and killed within minutes. But the iguanas flourish. They can do so because, unlike amphibians, their skin is watertight.

There are two kinds of iguanas on the islands – the land iguanas which live in the scrub, and marine iguanas that swarm on the bare lava fields by the coast. Basking in the sun, for them, is not a trial to be endured but, for most of the time, essential behaviour. The physiological processes of an animal's body, like all chemical reactions, are greatly affected by heat. Within limits, the higher the temperature, the quicker they proceed and the more energy they produce. Neither the reptiles nor the amphibians generate their heat internally; they draw it directly from their environment. Amphibians cannot expose themselves directly to the sun because of the permeability of their skins, so they must remain relatively cold and consequently sluggish. But the reptiles have no such problems.

The marine iguanas follow a daily routine that maintains their bodies at the most efficient temperature. At dawn they assemble on the tops of lava ridges or clamber onto the eastern faces of boulders, lying with their flanks broadside to the rising sun and absorbing as much heat as possible. Within an hour or so their temperature reaches its optimum level and they turn to face the sun. Now their flanks are almost in shadow and the rays strike only their chests. As the sun climbs higher and higher, the risk of overheating grows. Although reptile skin has the crucial quality of relative impermeability, it does not possess sweat glands, so the iguanas cannot cool themselves by allowing sweat to evaporate. Indeed, even if they could, this might not be a practical technique in an environment where water is so scarce. But they have to find some way of preventing themselves from simmering inside their skins.

Relief is hard to find. They stiffen their legs and hold their bodies off the baking black rock so that they absorb as little heat as possible from it, while what wind there is blows over their undersides as well as their backs. They pack themselves tightly into the few places where there is shade – in crevices beneath boulders or, better still, in the deep narrow gullies that are kept cool by the surging waves. The sea itself is too cold for comfort, for the Humboldt Current in which the Galapagos lie sweeps straight up from the Antarctic. The marine iguanas are compelled, however, to venture into it at some time every day to feed. Like many of their relations on the mainland of South America, they are vegetarians. No edible plants grow on the lava, but in the sea, just below high water mark, there are thick pastures of green algae. So, at some time during the middle of the day when their blood is almost as hot as they can tolerate and they are in danger of sunstroke, they risk a swim. They plunge into the surf, beating their tails like giant newts. Some hang on the rocks near the sea's edge, gnawing the seaweed with the sides of their mouths. Others swim farther out and dive to forage along the sea bottom.

Now their requirements are reversed. Instead of needing to disperse heat, they must retain it for as long as they can. They have a sophisticated physiological mechanism to help them: they can constrict the arteries near the surface of their bodies so that the blood, temporarily limited to the centre of the body, remains warmer longer. If they become too cold, they will lose the strength to swim back through the surf or to resist the tug of the waves as they cling to the boulders and, as a consequence, be smashed on the rocks. After a few minutes, that danger point has approached. The temperature of their bodies has dropped some ten degrees and they have to return to land.

Back on the rocks, they prostrate themselves, all four legs outstretched like a spread-eagled human bather exhausted after a chilling swim. Not until their body temperature has risen again will they be able to digest the meal that lies in their stomachs. As the sun starts to sink in the late afternoon, the risk of chilling returns and they assemble once more on the crests of the ridges to absorb as much as they can of the rays of the setting sun before the fall of night.

By such means, the iguanas manage to keep their bodies, for most of the time, very close to 37°C – almost exactly the temperature of the human body. Some lizards even maintain

The embryo must breathe, so the shell has to be slightly porous to enable oxygen to pass in and carbon dioxide to pass out. This shell brings complications, however, as well as benefits. Clearly, if it is dense enough to prevent the egg from drying out, then it is likely to prevent sperm getting in. Fertilisation must therefore take place within the female's body before the shell is deposited. To deal with this problem the male is equipped with a penis. The form of this organ varies considerably between the different groups of reptiles. Only one reptile today lacks such an organ, a strange lizard-like creature that lives on a few small islands in New Zealand, the tuatara.

The tuatara manages to achieve internal fertilisation in a manner that is reminiscent of some salamanders and frogs. When the pair come together, their genital openings are pressed closely together so that sperm from the male is able to actively swim into the female's oviduct. Interestingly enough, the tuatara has another characteristic reminiscent of amphibians. It is active at temperatures even below 7°C, which is much lower than any lizard or snake would favour. It seems therefore to be a very primitive kind of reptile, and the structure of its skull confirms this, for it resembles, in important ways, those of the earliest recognisable reptile fossils. Bones of a virtually identical creature have been found in rocks 200 million years old. The tuatara thus harks back, if not to the time when the reptiles first separated from the amphibians, then at least to an early stage in their history when, at the dawn of their golden age, the reptiles were beginning to diversify into a huge variety of forms.

At this time, the continents that we know today were united in one great landmass that geologists call Pangaea, and the four-legged, tough-skinned, egg-laying ectothermic reptiles now spread to all parts of it. The dry land became dominated by the dinosaurs, which ranged from creatures the size of chickens to monsters weighing over 30 tonnes. Others, the ichthyosaurs and plesiosaurs, took to the seas where their limbs became modified into paddles. And a third group, the pterosaurs, developed an expanse of skin on each arm stretching between one hugely elongated digit and the rest of their body that enabled them to fly. So the reptiles started their rule of the planet that was to last for the next 150 million years.

Some of the richest deposits of dinosaur remains lie in the midwestern states of North America. In Texas, the Paluxy River slowly meanders across a layer of mudstones that were once the mudflats of an estuary. One day at low tide several dinosaurs wandered across it. One was a theropod, a carnivorous species that walked erect on its hindlegs. The line of its three-toed footprints is still clear along one side of the present-day river. Farther down, the river has eroded more of the overlying rocks to expose in the same layer a trail of four immense circular prints nearly a metre across that were made by one of the huge plant-eating species. As the water ripples above them, it is easy to imagine that the river bed is not stone but still mud and that these giants were striding through the water only hours before.

At Dinosaur National Monument in Utah, a museum has been built around a cliff-face where a single layer of stone, some four metres thick, has yielded fourteen different species of dinosaur. Thirty complete skeletons have been taken away, but bones of many more remain. The rock which now forms the cliff-face was once a sandbank in the middle of a river. Gigantic rotting carcasses of dinosaurs floated down, beached on the sandbank and were dismembered there partly by putrefaction and partly by smaller dinosaurs that came to feast on carrion. All the long bones, such as those from the limbs and sections of backbone, lie pointing in roughly the same direction, and from them we can deduce which way the river ran. The whole deposit seems to have been laid down in a geological instant – perhaps over no more than a hundred years or so. It is an astonishing demonstration of how abundant those creatures once were.

Why did some species grow to such a great size? There are at least two possible reasons. The biggest, huge creatures with long necks and pillar-like legs, are known as sauropods. They grew to a length of around 25 metres and weighed perhaps 15 tonnes. Their teeth make it clear that they were vegetarians. The plants of the period – ferns and cycads – had tough fibrous fronds which would have certainly required a great deal of digestion. The teeth of sauropods, though very numerous, were simple and peg-like – totally incapable of grinding vegetation in the way that the molars of cows and antelopes do. The pulping of their food therefore had to be done in the stomach. Some species swallowed pebbles to act as millstones within their heaving stomachs, just as today, on a much smaller scale, some birds use grit in their gizzards. These stones, shiny from the joint effects of stomach acid and being ground against one another, are often found in a cluster between the ribs of herbivorous dinosaurs in exactly the place where their stomach once lay. The key digestive process however must have still been the biochemical power of the digestive juice and the bacteria that lived in their stomach. And that would have taken a considerable time. The herbivorous dinosaurs' stomach, therefore, had to be huge to serve as a storage vat where the food could be held while the lengthy process of fermentation took place. And a huge stomach requires a huge body to carry it.

The tree ferns and horsetails of the time grew much bigger than their present-day descendants. This may have been a consequence of competition to claim the light and overshadow competitors, or a way of avoiding being cropped by the large dinosaurs. At any rate, some of these plants grew to heights of six metres. Some vegetarian dinosaurs, however, grew long necks. This enabled them to reach more food than their competitors, and the fact that they did not need to chew it meant that their heads could remain small and slight. And this, in turn, allowed their necks to grow to even greater lengths. The carnivorous dinosaurs that preyed upon them would then also have to become giants in order to be able to subdue such huge prey. So the dinosaurs eventually became the largest animals ever to tread the lands of the earth.

The sauropods must have been relatively slow-moving, but the bones of many of the smaller dinosaurs make it clear that they were able, at least on occasion, to move very swiftly indeed. From that we can deduce that, at least at times, their blood temperature was quite high. Many were able to generate heat within their bodies. All endotherms today are equipped to conserve their warmth with some kind of heat insulation above or just below the surface of their skin – hair, fat or feathers. Without it the demands on their energy would be intolerable.

The fall of the dinosaurs' dynasty was catastrophic and virtually instant. Some 66 million years ago, a huge asteroid, perhaps as much as 15 kilometres across, crashed into the earth at a place called Chicxulub on the Yucatan peninsula of Mexico. The impact created tsunamis, firestorms, earthquakes and volcanic eruptions that buried some animals over 5,000 kilometres away. But the greatest effects were to come. The atmosphere had become thick with dust, blocking the warming rays of the sun. As a result, the world's climate cooled drastically and it remained so for perhaps a decade. The result was another of the great mass extinctions that have punctuated the history of life on earth. These geological, or, in this case, cosmological events radically changed the climate, and this, in turn, led to the extermination of vast numbers of organisms. Other events that occurred around the same time may also have played a role. There was a series of volcanic eruptions in what is now India. There may even have been other asteroid impacts. But the decisive event seems to have been the Chicxulub impact.

The dinosaurs were not the only victims. Perhaps 75 per cent of all animal species disappeared, including all the ammonites, as well as many sharks, some mammals, gigantic sea-going reptiles called mosasaurs, birds, lizards, and of course a great number of plants.

The rapidity with which the dinosaurs disappeared is visible with graphic clarity in the rocks of the Montana Badlands. Here, horizontal beds of sandstones and mudstones that were laid down 60 to 70 million years ago have been sliced and gouged by the violent storms of summer and the raging torrents flowing from the melting snows of winter into a wilderness of pinnacles, buttes and gullies. On the striped faces of the crumbling cliffs, trickles of brown fragments, like stains from a dripping tap, show where fossils are weathering out. Among them are the abundant remains of triceratops, a huge horned dinosaur. In life, this species grew to eight metres or so in length and may have weighed up to nine tonnes. Its immense skull carried three horns, one above each eye and one on the tip of its nose, as well as a great bony frill that projected from the back of its head. This certainly protected its neck, but it may also have been brightly coloured and could have been flaunted by the animals in aggressive displays. Its brain, however, was quite small – smaller certainly than that of a crocodile when the two are measured in the same way.

Just above the level at which the most recent of them are found, a thin deposit of coal rules a precise black line that can be traced from cliff to cliff across Montana and over the Canadian border into Alberta. This marks the death of the dinosaurs. Immediately below it, you can find the remains of not only triceratops but at least ten other species of dinosaur. Above it, there are none. One of the reasons why scientists came to accept the asteroid explanation of the extinction of the dinosaurs was that something like that layer is repeated all around the world and within it are found extraordinarily high levels of iridium, an element that is very rare on earth but is known to be present at high levels in asteroids. That thin line represents the blasted dust of the asteroid mixed with the rocks hurled from the crater created by the impact.

There were survivors of this cataclysm. Some mammals and reptiles escaped. So did many amphibians and birds. The reptiles that survived while their larger relatives died may have found two ways to escape the effects of a drop in temperature. Both methods are practised by various reptiles alive today. One is to find a crevice in rocks or to bury yourself so that you are beyond the reach of the worst frosts, and then to fall into a state of suspended animation – to hibernate. The other is to take to the water. Water retains its heat much longer than air, so the effects of decades-long winter are much reduced. It may also be that some species escaped by swimming to warmer latitudes. Significantly, the three main types of reptile that survive today from the time of the dinosaurs – the crocodiles, the lizards and the tortoises and turtles – are able to take advantage of one or other of these expedients.

The crocodiles are the largest of all living reptiles. Males of the huge sea-going species that lives in Southeast Asia have been reported to be over six metres long. Fossils of them first appear in the rocks at about the same time as dinosaurs, and species very like the monsters of today lived alongside sauropods and doubtless preyed on smaller antelope-sized ones. If anyone supposes that this dinosaur-ruled world was one of puny-brained animals lumbering clumsily about, reacting in a simple, slow-witted way to one another, then watching crocodiles today will quickly show how false that picture must be.

The Nile crocodile spends most of its days basking on sandbanks, maintaining an even body temperature in much the same way as the Galapagos iguanas. Its problem, however, is not as acute as the iguanas', since, being so much bigger, it is less affected by short-term variations. It also makes particular use of an additional technique for cooling. It opens its mouth and gapes so that air plays over the soft skin inside the mouth, which is much thinner than the hide that covers the body. At night, when even in the tropics the temperature can fall quite low, it moves down into the warm waters of the river. Although crocodiles are inactive for long periods, on occasion they can run very fast indeed. Their social lives are quite complex. The males establish a breeding territory, patrolling a patch of water not far from a beach. They bellow and fight any other males that come to dispute with them. Courtship takes place in the water. As the female approaches, the male becomes greatly excited. His roars increase to such an intensity that his flanks vibrate and throw up clouds of spray. He lashes his tail and claps his huge jaws. Actual mating lasts only a couple of minutes or so. The male clasps the female with his jaws and their tails intertwine.

The female digs a hole well above the waterline on a site that she may use all her life. She lays at night, producing 40 eggs in several batches. The depth to which she buries them varies according to the nature of the soil, but it is always sufficiently deep for the temperature not to vary by more than 3°C. The holes are never made in places that are exposed to full sunshine throughout the day. Some species go to even greater lengths to ensure that their eggs remain at an even temperature. The saltwater crocodile builds a mound of vegetation as a nest and sprays urine over it when the heat gets too intense. The American alligator also piles up vegetation, lays her eggs in it, and regularly turns it over to provide the eggs beneath with moisture and a constant heat from the rotting foliage. One of the reasons they take such care to maintain a stable temperature is that in crocodilians, as in some other reptiles and fish, the sex of an animal is determined when still in the egg by the temperature at which it develops. At higher temperatures, a greater proportion of young will be female when they hatch.

It is in the care it gives its offspring that the crocodile's behaviour is most complex and surprising. When the eggs of the Nile crocodile are close to hatching, the young within begin to make piping calls. These are so loud that they can be heard, through shell and sand, from several metres away. In response, the female begins to scrape away the sand covering the eggs. As the young struggle up through the sand, she picks them up with her jaws, using her huge teeth as gently and delicately as forceps. A special pouch has developed in the bottom of her mouth and in it she can accommodate half a dozen babies. When she has collected such a number, she carries them down to the water and swims away, with her jaws half closed, her young passengers piping and peering through the palisade of teeth. The male helps, and within a short time the young have been ferried to a special nursery area in the swamp. Here they remain for a couple of months, hiding in small holes in the bank and hunting for frogs and fish while their parents laze in the water close by, keeping guard. Watching them, one can well believe that the dinosaurs themselves had similarly complicated forms of courtship and parental behaviour.

The tortoises have an ancestry just as ancient as the crocodiles. Very early in their history, they invested in defence. The crocodiles had strengthened their skin with small ossicles beneath the scutes of their backs. The tortoises took even more extreme measures, enlarging the scales into horny plates. They also reinforced them from below by expanding and flattening their ribs to create a continuous bony shield just beneath the skin. As a result, their bodies became enclosed within a virtually impregnable box into which they could withdraw their head and limbs when danger threatened. But this had serious consequences. Many reptiles and indeed mammals like ourselves draw air into our lungs by expanding our chests. Lifting our ribs enables us to do this. But the tortoises with their ribs flattened and joined together cannot do so. Instead, they have had to develop another method. They use a unique kind of muscular sling that creates an internal diaphragm which inflates and deflates the lungs. It may not be quite as efficient a way of breathing, but the tortoises now have by far the most effective armour developed by any vertebrate. It has certainly served the tortoises well, for they have remained virtually unchanged from that day to this.

The one major variation on the basic pattern arose very early in their history. One group took to the water and became the turtles. It was a logical move for a creature with heavy bulky armour that made movement on land laborious and energy-consuming, but one of their newly acquired reptilian talents prevented them from becoming totally at home there. The shelled egg that had enabled their ancestors to become independent of water was useless in it. In water, the young would drown within their shells. So the female turtle, every breeding season, has to forsake the open ocean, swim to coastal waters and then, one night, haul herself with great labour up a sandy beach, excavate a hole and lay eggs, just as her land-living relations do.

The third group of reptilian survivors, the lizards, are now very much more numerous than either the crocodiles or the tortoises and turtles. They have also changed to a much greater degree from their ancestral pattern. There are now many different families – iguanas, chameleons, skinks, monitors and several others. They have all protected their invaluable watertight skin by modifying their scales. The Australian shingleback skink has a covering of stout polished ones that fit together with the neatness of chain mail; the Gila monster from Mexico is clothed in rounded black and pink ones resembling beads; the African sun-gazer grows them long and spiny like rococo armour.

Scales, like our own fingernails, are made of a dead horny material and gradually wear away. The lizards have therefore to replace them, often several times a year. A new set grows beneath the old, which is then sloughed off.

Scales are, it seems, more quickly responsive to evolutionary pressures than bone, and they serve the lizards in many ways apart from straightforward protection against wear and tear. The marine iguanas have a crest of long ones along their spine so that the males, when they display in territorial competition, appear particularly big and formidable. Some chameleons, the most heraldically dramatic of all reptiles, have grown their head scales into horns – single, double, triple or even quadruple. The thorny devil, a tiny, highly specialised lizard from the central Australian desert which lives entirely on ants, has each scale enlarged and drawn out to a point in the centre. Few birds could relish such a thorny mouthful and they must be a very effective defence, but the shape of the scales also serves another and most unusual function. Each is scored with very thin grooves radiating from the central peak. During cold nights, dew condenses on them and is drawn by capillary action along the grooves and eventually down to the tiny creature's mouth.

Perhaps the most specialised scale of all is that developed by the geckos. These small tropical lizards can run up walls, scuttle upside down over ceilings, even cling to vertical panes of glass – and do all these tricks with such ease that it is tempting to think that suction is involved in some way. But scales are responsible. Those on the underside of the toes carry pads formed from enormous numbers of microscopic hairs, invisible to the naked eye. Each hair is so tiny that it can only be seen through the electron microscope. When the toes are pressed down, electromagnetic forces bind those hairs to a surface, and this gives the gecko a foothold on even the smoothest surface, including glass. To break the attachment, the gecko changes the angle of the hairs' contact, and this enables it to lift its feet.

Throughout their history, the lizards, like the salamanders of the New World, seem to have had an evolutionary tendency to lose their legs. Several skinks today represent different stages of the process. Australian ones, like the blue-tongue or the shingle-back, have, at best, diminutive legs which are scarcely sufficiently strong or big to hoist their stout bodies above the ground. The European slow-worm, another lizard, has no legs at all, though internally it still carries relics of shoulder and hip bones. The snake-lizards of South Africa, even within their single genus, show many intermediate stages of limb reduction. One has all four legs, each with five toes; the limbs of another are very small and have only two fully developed toes; and a third has hindlegs with one toe apiece and no external front legs at all.

A hundred million years ago, this process of limb reduction took place among a group of early lizards. Its consequence was the appearance of the snakes. The exact identity of this ancestral group is still in debate. The loss of their limbs, however, seems to have been connected with the assumption of a burrowing life. There are several clues that suggest that the snake's ancestors once lived underground. Burrowing could easily damage the delicate drum of the ear, and hearing is not, in any case, of much value. So burrowers tend to lose their ears. No snake has an eardrum, and the bone that in other reptiles transmits vibrations from an eardrum is connected instead to the lower jaw. As a consequence, snakes are virtually deaf to sounds transmitted through the air but can, instead, detect vibrations, such as those produced by a footstep, that travel through the ground.

Their eyes, too, provide further evidence, for they differ considerably in structure from any other reptilian eye. If the snake's ancestors had been burrowers, then their eyes, like those of any other burrower, would have tended to degenerate. But if, before they were lost altogether, their owners had returned to a life above ground, then sight would once again be needed and the vestiges would redevelop. So the snake eye would have a structure peculiar to itself. This explanation is very persuasive but not yet universally accepted.

No one doubts, however, that the snakes once had legs. Indeed, a whole group of them, the pythons and boas, still retain internal relics of their hip bones, and show external signs of them – two spurs on either side of the vent. Above ground, without legs, the snakes had to develop new means of getting about. They flex their flank muscles in alternate bands so that their body is drawn up into a series of S-shaped curves. As the contractions travel in waves down the body the flanks are pressed against objects on the ground such as stones or plant stems and the snake is able to push itself forward. In short, it wriggles. If it is put on a surface completely free of any irregularities to provide purchase, the technique fails and the snake simply writhes helplessly.

Several snakes that live in sandy deserts have developed a variation of this technique and practise it so quickly that it is baffling to watch and extremely difficult to describe comprehensibly. It is called side-winding. The snake contracts its body into an S-shape as other snakes do, but it only touches the ground at two points, which travel rapidly down the body. The movement starts behind the head. The snake lifts its head and bends it into a curve at the point where it touches the ground. The muscle contraction making the bend travels rapidly down the body, keeping contact with the sand beneath while the forepart and head remain raised. By the time the wave is halfway down, the neck drops once more and momentarily touches the sand as a new wave begins again. The result is that the snake moves rapidly forward, leaving behind a series of bar-like tracks in the sand, orientated at about 45° to the direction in which the snake has actually travelled.

When a snake hunts, it is often very important for it to be able to advance with a minimum of movement so as not to attract the attention of its victim. The snake lies with its body quite straight and pointing directly at its prey. The scales on its underside are shaped like narrow rectangles running across the width of the body and overlapping one another with their free edges to the rear. The snake is able to hitch these scales up and forward in groups by contracting its belly muscles. The back edges catch on the ground, and as the contractions pass downwards in waves, the snake advances smoothly and silently with no lateral movement whatsoever.

If the ancestral snakes did indeed spend a period below ground, their prey is likely to have been small and limited to invertebrates such as worms and termites and perhaps the early burrow-living shrew-like mammals. When they came above ground, after the mammals had begun to evolve into the forms we know today, their scope became very much greater. There is clear evidence that the range and variety of the snakes increased enormously after the disappearance of the dinosaurs and the rise of the mammals and birds. A few boas and pythons now grow to such a length that they can tackle creatures as big as goats and antelope. Having seized their prey with their mouths, they swiftly coil themselves around it and then kill it by tightening their coils so that their victim cannot expand its chest to breathe. It dies by suffocation rather than crushing. With the backward-pointing teeth engaging on the prey, the snake draws in its food by working its loosely connected jaw. The long process of swallowing may take several hours and, once achieved, often leaves the snake in a state of bloated immobility.

The more advanced snakes kill, not by constriction, but by poison. One group, the back-fanged snakes, delivers the venom by means of specially adapted teeth near the back of the upper jaw. The poison glands lie above these teeth and the venom simply trickles down a groove in the tooth. Once the prey has been bitten, the back-fanged snake may have to maintain its grip and chew, rocking its jaw from side to side until the fangs at the rear of the mouth are at last driven into the victim's flesh, delivering the poison.

Some snakes have an even more refined way of killing. Their fangs are placed at the front of the upper jaw and have an enclosed canal through which the venom flows. Cobras, mambas and sea-snakes have fangs that are short and immobile, but those of vipers are so long that most of the time they have to be kept hinged back, lying flat along the roof of the mouth. When the snake strikes, its mouth opens wide, and the bone to which the fangs are attached rotates, bringing the fangs down and forward so that they will stab the victim immediately. When they pierce the flesh, the venom is injected down them, like serum from a hypodermic needle.

The snakes were the last of the great reptile groups to appear, and the most sophisticated of them are the pit vipers. The rattlesnakes of Mexico and the southwest of the United States belong to this group and exemplify the perfection to which the reptilian pattern has been taken.

Like many other snakes, and some amphibians and fish before them, the rattlesnakes give their eggs maximum protection by retaining them inside the body. That reptilian innovation, the shell, is reduced to a thin membrane so that the embryos, as they lie inside the oviduct, not only feed on their yolk but draw sustenance from their mother's blood diffusing from the walls of the oviduct pressed against them. It is a process which parallels, in essence, the device of the placenta that is used by the mammals.

Nor does the female rattler abandon her young once they emerge fully formed from her vent. She actively guards them. Intruders are warned off with the sound of her vibrating rattle. Each time she sheds her skin, one special hollow scale remains fixed to her tail, so that a fully grown rattler may have as many as twenty of them.

A rattlesnake hunts mostly at night and does so with the aid of a sensory device which has no parallel elsewhere in the animal world. Between the nostril and the eye is the pit which gives the group as a whole its name. It detects infrared radiation, that is to say heat, and the cells lining its internal surface are so sensitive that they can detect a rise of three hundredths of one degree centigrade. What is more, it is directional so that the snake is able to locate the source of the heat with precision. So, with the aid of its pits, the rattlesnake is able to detect the presence of a small ground squirrel crouching motionless half a metre away. The snake glides smoothly towards it on its belly scales in near-silence; once within range, it strikes, shooting its head forward at a speed of three metres a second; and then its huge paired fangs inject its victim with a dose of extremely virulent venom. It must surely be one of the most efficient killers in the animal world.

Because, like all reptiles, it can absorb the sun's energy directly, a rattlesnake's food requirements are small. A dozen or so meals a year are quite sufficient for it. Not for the rattlesnake the incessant daily search for food to which the endothermic mammals, even in a desert, are committed. Nor does it need, like them, to spend its days cowering in crevices and holes, panting with the heat, waiting for the cool night to fall before it can venture abroad. Curled up among the stones and cactus of the Mexican desert, a rattlesnake is the master of its environment and fears nothing. The reptiles, by virtue of their watertight skins and eggs, were the first vertebrates to colonise the desert. In some places, some of them still own it.

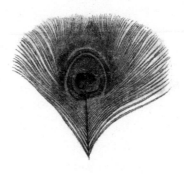

EIGHT

Lords of the Air

The feather is an extraordinary device. Few substances can equal it as an insulator and none, weight for weight, whether man-made or animal-grown, can excel it as an aerofoil. Its substance is keratin. The same horny material forms a reptile's scales and our own nails, but the exceptional qualities of a feather come from its intricate construction. A central shaft carries on either side a hundred or so filaments; each filament is similarly fringed with about a hundred smaller filaments or barbules. In downy feathers, this structure produces a soft, air-trapping fluffiness and, therefore, superb insulation. Flight feathers have an additional feature. Their barbules overlap those of neighbouring filaments and hook them onto one another so that they are united into a continuous vane. There are several hundred such hooks on a single barbule, a million or so in a single feather; and a bird the size of a swan has about 25,000 feathers. Almost all the characteristics that distinguish birds from other animals can be traced one way or another to the benefits brought by feathers. Indeed, in the modern world, the very possession of a feather is enough to define a creature as a bird.

When, in 1860, in Solnhofen in Bavaria, the delicate and unmistakable outline of a single isolated feather, seven centimetres long, was found impressed in a slab of limestone, it caused a sensation. It lay on the rock, as eloquent as a native American sign, proclaiming that a bird had been there. Yet these limestones dated from the days of the dinosaurs, long before birds were thought to exist.

The sediments from which they are formed were deposited on the bottom of a shallow tropical lagoon enclosed by a reef of sponges and lime-depositing algae. The water was tepid and poor in oxygen. Cut off from the open sea, there were few if any currents. Limy mud, partly from the disintegrating reef and partly produced by bacteria, was being deposited as ooze on the bottom. Such conditions suited few animals. Those that did stray there died, fell to the bottom and lay undisturbed in the still water as they were covered by the slowly accumulating ooze.

The Solnhofen limestones have been quarried for centuries because their fine, even grain makes them excellent for building. In the nineteenth century, they proved to be ideal for carrying images in the newly invented printing technique of lithography. And they are also immaculate blanks for nature to impress with the fine detail of the evidence of evolution.

The stone, if it is thoroughly weathered, splits along the bedding planes so that a block can be opened into leaves, like a book. When you visit one of the quarries, it is almost impossible to resist the temptation to turn the pages of every boulder that you see, knowing that no one has ever done so before and that whatever they contain will not have been exposed to daylight for 140 million years. Most, of course, are blank, but every now and then the quarrymen find fossils of a near-miraculous perfection – fish with every bone and shining scale in place, horseshoe crabs lying just where they died at the end of their last trail through the silt, lobsters with even their finest antennae intact, small dinosaurs, ichthyosaurs and pterodactyls, lying with the bony scaffolds of their wings crumpled but unbroken and the shadow of their leathery flight membranes plain to see. But in 1860, that beautiful and enigmatic feather was the first indication that birds had been living in such company.

To what kind of bird had it belonged? Science, on the strength of the feather alone, called it Archaeopteryx, 'ancient feather'. A year later, in a quarry close by the first, someone discovered an almost complete skeleton of a feathered creature the size of a pigeon. It lay sprawling on the rock, its wings outstretched, one long leg disarticulated, the other still connected with four clawed toes, and all around it, dramatically and indisputably, the clear impress of its feathers. It certainly seemed apt to call it an 'ancient bird', but it differed substantially from any known living bird. The long feathered tail that flared out behind it was supported by a bony extension of its spine; and it had claws, not only on its feet but on the three digits of its feathered forelimbs. It was almost as much a reptile as a bird, and its discovery within two years of the publication of *The Origin of Species* was a providentially timed confirmation of Darwin's proposition that one group of animals developed into another by way of intermediate forms. Indeed, Thomas Huxley, Darwin's champion, had predicted that just such a creature must have existed, and had prophetically described its details. Even today, there is no more convincing example of such a link.

Since that first specimen was discovered, at least nine more have been found in the Solnhofen rocks, so today we have extremely detailed knowledge of this extraordinary animal's anatomy. Some of its characteristics are clearly reptilian: a skull with bony jaws studded with teeth, hindlimbs and forelimbs each with separate digits armed with sharp curved claws, and a long bony rod in its tail. But it also has feathers that clothe its entire body. Those on its forelegs are particularly long and broad and clearly capable of catching the air sufficiently effectively to enable it to fly.

One bird alive today shows how useful that combination of claw and feather can be for a climber. The hoatzin is a curious, heavily built bird, the size of a chicken, that lives in the swamps of Guyana and Venezuela. Its nests are roughly constructed platforms of twigs, built above water, often in mangroves. When the young first hatch, they are naked and extremely active. Watching them is not easy. If a bump from the bow of your canoe shakes the mangrove branches, the young will scramble feverishly off their twig platform. But they have two tiny claws on the front edge of each wing which enable them to cling to the twigs around them. If they are disturbed further, they will abruptly launch themselves into the air, dive in the water and swim energetically into the tangle of mangroves where you will never be able to follow them. But, watching them, you may get a hint of the way that Archaeopteryx must have moved through the branches of its dinosaur-haunted forests.

But Archaeopteryx's feathers had never been seen before on such an ancient creature. How had it acquired them? Did its ancestors clamber in to the branches using the claws on its wings and then start to glide from branch to branch? Small lizards do that in Borneo, using flaps of skin extending from their flanks and held taut by extensions of their ribs. Or perhaps their ancestors were ground-living and acquired feathers on their forelegs to help them sweep up insects for food. Did they perhaps help them to escape from predators by enabling them to make sudden leaps into the air. The arguments were vigorous and even heated, until in the 1980s, new and almost unbelievable evidence came from China. Skeletons of dinosaurs were discovered that were clearly ground-living but nonetheless covered in feathers.

They came from Liaoning province in the north-east of the country and lay in mudstones and shales that had formed on the floor of vast lakes. Some belonged to small, flesh-eating dinosaurs called theropods. The group was already well known. Near-perfect skeletons had been excavated elsewhere. But only in the Liaoning deposits were conditions for fossilisation so perfect that they had preserved traces of feathers. These feathers covered the entire body.

Their primary function, clearly, was to insulate the body and so maintain the warmth needed for high activity.

So feathers did not evolve initially for flight and do not define birds. The birds inherited them from feathered ancestors. And this leads to the realisation that all dinosaurs were not exterminated when the asteroid crashed into Mexico 66 million years ago. Some of those with feathers survived. So dinosaurs today are flying around our gardens.

But evolution would have to bring many profound changes before those little feathered dinosaurs became the extremely efficient flying creatures we know today. The overwhelming evolutionary pressure came from the advantages of reducing weight. Archaeopteryx's bones were solid, like a modern reptile's. Those of true birds are paper-thin or hollow, often supported inside with cross-struts which closely resemble those designed to strengthen the wings of aeroplanes. Birds' lungs are extended into air sacs which bulge into the body cavity, so filling space in the lightest possible way. The heavy bony extension of the spine that formed the basis of Archaeopteryx's tail has been replaced with stout-quilled feathers requiring no bony support of any kind. A weighty jaw, laden with teeth, must have been a particular handicap for any creature trying to fly, for it would tend to unbalance the animal and make it very nose-heavy. Modern birds have lost it and have developed instead another lightweight construction of keratin, the beak.

But even the best beak cannot chew, and most birds still have a need to break up their food. They do so with a special muscular compartment of the stomach, the gizzard, as some of their distant sauropod ancestors did. So the beak itself has to do no more than gather the food.

The keratin of the beak, like that of the reptilian scale, seems to be easily moulded by evolutionary pressures. Just how quickly it can be changed to suit the diet of its owner is vividly shown by the honey creepers of Hawaii. The ancestor of these birds was probably a finch with a short, straight beak that lived in continental America. A few thousand years ago, a flock of them must have been carried out to sea by a freak storm. They eventually reached the Hawaiian islands, and there found lush forests empty of other birds, for the islands are volcanic and were formed comparatively recently. To exploit the many kinds of food now at their disposal, they rapidly evolved into over 50 different species, each specialised for a particular diet with the beak shape that was best suited to gather it. Some have short thick bills for seed-eating, others have hooked and powerful ones for tearing carrion. One species has a long, curving bill for extracting nectar from lobelia blossoms; another has an upper mandible twice the length of the lower, which it uses to hammer bark and lever it off in its search for weevils; yet another has crossed mandibles, a form that apparently enables it to extract insects from buds. Darwin had noted similar variations in the bills of the finches of the Galapagos Islands and regarded them as powerful evidence for his theory of natural selection. He never had the luck to visit Hawaii. Had he done so, he might well have concluded that the honey creepers illustrated his arguments even more convincingly.

Elsewhere in the bird world, where the evolution of

beaks to suit a particular purpose has gone on for much longer, there are even more extreme forms. The sword-billed hummingbird has a probing beak four times the length of its body with which it sucks nectar from deep-throated Andean flowers. The macaw has a hooked nut-cracker of such strength that it can split that most intractable of nuts, the brazil nut. The woodpecker uses its beak like a drill to excavate wood-boring beetles. The flamingo's crooked beak has inside it a fine sieve through which it pumps water with its throat and so collects tiny crustaceans. The skimmer has a lower mandible almost twice the length of the upper so that it can fly low over a river with the lower mandible just cutting the surface of the water. When it touches a small fish the bill snaps shut instantaneously and the fish is caught. The list of odd bills is virtually endless and ample proof of the malleability of the keratin beak.

Significantly, most of these foods – fish, nuts, nectar, insect larvae, sugar-laden fruit – are full of calories. Birds favour them because flying is an extremely energetic business. To ensure that energy in the form of heat is not wasted, insulation is of the greatest importance. So feathers are essential to a bird not only to provide aerofoils on the wings but to enable it to generate enough energy to flap them.

As insulators, feathers are even more efficient than fur. Only a bird – the emperor penguin – can survive on the Antarctic icecap in winter, the coldest place on earth. Penguin feathers are devoted entirely to this task. They are filamentous and trap the air in a continuous layer all round the body. This, reinforced by a thick coat of fat just beneath the skin, enables the warm-blooded emperor penguins to stand about in a blizzard in temperatures of forty degrees below freezing and remain there for weeks on end, even without stoking their internal warmth with a meal. And when we go there, the most luxurious and effective way we have yet devised of keeping our own bodies warm is with feathers taken from an Arctic duck – eiderdown.

The feathers upon which a bird's life is so dependent are regularly moulted and renewed, usually once a year. Even so, they need constant care and servicing. Their owners wash them in water and ruffle them in dust. Disarranged feathers are carefully repositioned. Those that have become bedraggled or have broken vanes are renovated by careful combing with the beak. As the filaments pass through the mandibles and are pressed together, the hooks on the barbules re-engagelike teeth of a zip-fastener to make a smooth and continuous surface again.

Most birds have a large oil gland in the skin near the base of the tail. The bird takes the oil from it with its beak and anoints its feathers individually so that they are kept supple and water-repellent. Some birds, including herons, parrots and toucans, lack this gland. They condition their feathers with a fine talc-like dust, powder-down, that is produced by the continuous fraying of the tips of special feathers which grow sometimes in a clump, or are scattered through the plumage. Cormorants and their relatives the darters, although they spend a great deal of their time diving in water, have feathers so constructed that they get thoroughly wet, but this is to their advantage, for by losing the air trapped beneath them, they become much less buoyant and so can dive in pursuit of fish with greater ease. When they have finished fishing, they have to stand on the rocks, wings outstretched, drying themselves.

The skin beneath the feathers must be an extremely attractive place for fleas, lice and other parasites. There, they are warm, snug and out of sight. Many such creatures may afflict a bird, so birds regularly erect their feathers and probe around the base of their quills to pick off lodgers. Jays, starlings and jackdaws and several other species actively encourage insects to crawl over their skin, probably as an aid in this de-lousing process. The bird will squat on an ant nest with feathers ruffled and spread, so that the disturbed, angry ants swarm all over it. Sometimes it even picks up individual ants with its bill, holding them firmly but gently so that they are not killed, and then the bird jabs its skin and strokes its feathers with them. The ants usually chosen for this are those which eject formic acid when irritated, and this would undoubtedly kill parasites. The behaviour may have originated as a matter of personal hygiene, but now some individual birds seem to do it for pleasure and will 'ant' with all kinds of things that might give their skins exciting and pleasurable stimulation – wasps, beetles, smoke from a fire, even lighted cigarette ends. Anting sessions may go on for half an hour or so, the bird sometimes falling over itself excitedly in its attempts to stimulate parts of its body that are difficult to reach.

All this toiletry takes up a considerable part of the bird's non-flying time. The reward comes when it takes to the air. The immaculately arranged feathers not only form perfect aerofoils on the wings and tail, but those on the head and the body fulfil the equally valuable function of streamlining the contours so that there is a minimum of eddying and drag when the bird is in flight.

Bird wings have a much more complex job to do than the wings of an aeroplane, for in addition to supporting the bird they must act as its engine, rowing it through the air. Even so, the wing outline of a bird conforms to the same aerodynamic principles as those eventually discovered by humans when designing aeroplanes, and if you know how different kinds of aircraft perform, you can predict the flight capabilities of birds with similar profiles.

Short stubby wings enable a tanager and other forest-living birds to swerve and dodge at speed through the undergrowth just as they helped the fighter planes of the Second World War to make tight turns and aerobatic manoeuvres in a dogfight. More modern fighters achieve greater speeds with swept back wings, just as peregrines do when they go into a 130 kph dive, stooping to a kill. Gliders have long thin wings so that, having gained height in a thermal up-current,they can soar gently down for hours, and an albatross, the largest of flying birds, with a similar wing shape and a span of 3 metres, can patrol the ocean for hours in the same way without a single wingbeat. Vultures and hawks circle at very slow speeds supported by a thermal, and they have the broad rectangular wings that very slow flying aircraft have. Designing wings to provide hovering flight has proved beyond us – the only way we have found to do this is with the whirling horizontal blades of a helicopter or the downward-pointing engines of a vertical landing jet. Hummingbirds used this technique long before us. They tilt their bodies so that they are almost upright and then beat their wings as fast as 80 times a second, producing a similar down-draught of air. So the hummingbird can hover and even fly backwards.

No other creatures can fly as far, as fast or for as long as birds. The swift is, indeed, the swiftest – one Asian species being capable of speeds of 170 kph in level flight and flying every day about 900 kilometres to collect the insects that are its only food. So extreme is its adaptation to an aerial existence that its feet are reduced to little more than tiny grasping hooks. Its scimitar-curved wings are so long that sitting flat on the ground it cannot beat them properly and it can only get into the air with any ease by launching itself from a cliff or the side of its nest. It even mates in mid-air. A female, flying high, holds out her wings stiffly and a male comes from behind, alights on her back and, for a few moments, the two glide together. They never alight between breeding seasons so that they spend at least nine months of the year continuously on the wing. Even that, however, is excelled by the sooty tern, which, after it leaves its nest for the first time, has not been seen to alight or settle on the water until it nests three or four years later.

Many species of birds make long annual journeys. The European stork travels every autumn down to Africa and returns to Europe in the spring navigating with such accuracy that the same pair, year after year, will return to the same roof top to occupy the same nest. The greatest traveller of all is the Arctic tern. Some nest well north of the Arctic Circle. A chick hatching in northern Greenland during July will, within a few weeks, set off on an 18,000-kilometre flight that takes it south, down the western coasts of Europe and Africa and then across the Antarctic Ocean to its summer grounds on the pack ice not far from the South Pole. It may then, during the Antarctic summer, driven by the incessant westerly gales, circle the entire Antarctic continent before leaving once more for southern Africa and then heading north back to Greenland. So it experiences both the Antarctic and the Arctic summers when the sun scarcely dips below the horizon, and sees more daylight each year than any other creature.

The energy spent by such migrants in their vast journeys is gigantic, but the advantages are clear. At each end of their routes they can tap a rich food supply that exists for only half the year. But how did they ever discover that such sources existed so far apart? The answer seems to be that their journeys were not always so long. It was the warming of the world at the end of the Ice Age 11,000 years ago that began to stretch them. Before that time, birds in Africa, for example, might fly briefly a little to the north to the edge of the icecap in southern Europe where, for a few months in summer, there were insects in quantity and no permanent local population to feed on them. As the glaciers began to retreat, new strips of land became liberated from the ice and colonised by insects and berry-bearing plants. So each year, birds were able to find food by flying farther and farther until their annual journeys involved travelling thousands of kilometres. Similar climatic changes are likely to have been responsible for extending the movements of those migrants in Europe and North America which fly in an east–west direction to the centre of the continents during the summer and back for the winter to the coastal regions that are kept warmer by the sea. Such changes are still taking place. The blackcap, which in summer months lives in Germany, now overwinters in the UK as well as in its traditional site, Spain. As a result, the birds that migrate to the UK are becoming increasingly different from the Spanish ones. They may be on the way to forming two separate species.

But how do the birds manage to find their way? There seems to be no single answer: they use many methods. Many birds certainly follow major geographical features. Summer migrants from Africa fly along the North African coast, converging on the Strait of Gibraltar, and cross there, where they can see Europe ahead of them. Then they follow valleys, flying over recognised passes through the Alps or the Pyrenees, and so arrive at their summer homes. Others take an eastern route by way of the Bosphorus.

But all birds cannot use such straightforward methods. The Arctic tern, for example, has to fly at least 3,000 kilometres across the Antarctic Ocean with no land to guide it. We know that some birds, flying at night, navigate by the stars, for on cloudy nights they tend to get lost, and if they are released in a planetarium where the constellations have been rotated so that they no longer match the position of the stars in the heavens, the birds will follow the visible and artificial ones.

Day-flying birds may use the sun. If they are to do so, they must be able to compensate for the shift of the sun across the sky each day, and this means that they must have a precise sense of time. Still others appear to be able to use the earth's magnetic field as a guide. So it seems that many migrating birds must carry in their brains a clock, a compass and the memory of a map. Certainly, human navigators would need all three if they were to match the journeys that a swallow can make within a few weeks of its hatching. One sense that many birds use that a human would find much more difficult to match is their sense of smell. Many migrating birds will use local odours to enable them to pinpoint a site where they have previously been, as they come in to make their final approach.

Even when we know what abilities birds are using during their migrations, their skills can be astonishing. In a famous case, a shearwater was taken from its nest on the island of Skokholm in west Wales and sent by aircraft to Boston in the United States, 5,100 kilometres away. There it was released. It was back in its breeding burrow twelve and a half days later, a time so short that the bird must have flown in a direct and purposeful way. Simply extraordinary.

The feathers that keep a bird warm and enable it to fly are of service to it in yet a third way. Their broad surfaces, easily erected or folded away, serve splendidly as banners with which to send messages. This, indeed, may also have been one of the ways in which non-avian dinosaurs used them. For most of their lives the majority of birds have much to gain from remaining inconspicuous, and feathers can provide the colours and patterns necessary for perfect camouflage. But each year, at the beginning of the breeding season, birds have an overriding need to communicate with one another. As male meets male in territorial dispute over nesting sites, dramatic feather crests are raised, coloured chests pouted and wing patterns spread, in a long series of ritualised threats, arguments and appeals. These visual signals are usually reinforced with vocal proclamations. Both kinds of signals carry the same three messages – a declaration of species; a challenge to any male of that species to dispute the ownership of the territory; and an invitation to a female to join him.

The nature of the country a male bird inhabits and his general character may make one medium of communication more suitable than the other. Shy birds that normally live unobtrusive lives in woodland or thick forest tend to use only the minimum of visual signals and concentrate instead on pouring out a specially long and elaborate song. If you hear marvellous cascades of notes, full of liquid trills and thrilling swoops, the singer is likely to be a plain-liveried unspectacular creature – a bulbul in Africa, a babbler in Asia, a nightingale in Europe. Conversely, the most gorgeously caparisoned birds – peacocks, pheasants, parrots – are those that are so self-confident, so untroubled by the fear of enemies, that they have no hesitation in exposing themselves in prominent places to show off their adornments. Since their main signal is a visual one it is no surprise that such birds usually have calls that are short, uncomplicated and harsh. We now know that elaborate song evolved three times in separate bird lineages in the last 30 million years or so. This means that the earliest birds as well as the feathered dinosaurs would not have sung sweetly but would – at best – have croaked or growled or hissed.

A declaration of the species of the sender is obviously important to prevent birds wasting their time in courting and coupling with partners with whom there can be no fertile union. In a few cases, this is done entirely by song. A human ornithologist and a female bird may be equally baffled about the identity of a small brown warbler lurking in an English hedgerow. Neither may be sure exactly who he is, judging solely from his appearance. It is only when he begins to sing that either can tell whether he is a willow warbler, a wood warbler or a chiffchaff.

Usually, however, identity of species is proclaimed by the plumage, a fact that a heartless experimenter can demonstrate by painting an eye-stripe or a wing-flash on a bird so that it looks like a related species and successfully deceives a genuine member of it. Identification becomes a particular problem when many related species live in the same area and there is danger of confusion between them. This was the problem that produced the brilliant and varied colours of the closely related butterfly fish on the coral reef. Similarly, if extravagant feather patterns and bright colours are found in many closely related birds it may well be an indication that these birds often share the same habitat. Some of the most vividly coloured birds in Australia are the parakeets and finches. In both groups, several species do indeed live in the same patch of country. All over the world, ducks of different species mingle together in large assemblages on open water in spring. The drakes of each species develop for the occasion highly characteristic patterns and colours on their heads and wings so that the females can recognise them. That the prevention of confusion is a major function of these colours is shown by the fact that when only one species of duck manages to colonise an island, and remains there long enough to develop into an individual form, it is always a much drabber version of its mainland original. There is no longer any need for such a drake to send vivid signals about who he is: there is no other bird around with whom his females could confuse him.

At the same time as proclaiming their species, individual birds must also declare their sex to one another. Ducks do so with their head patterns, for only the drakes develop them. In many species, however – among them seabirds and birds of prey – the male and female look the same throughout the year. Their sexual identity therefore has to be conveyed by their song and behaviour. The male penguin has a particularly charming way of discovering what he wants to know about his uniformly suited companions. He picks up a pebble in his bill, waddles over to a bird standing alone and solemnly lays it before it. If he gets an outraged peck and the squaring up for a fight, he knows he has made a dreadful mistake – this is another male. If his offering is met with total indifference, then he has found a female who is not yet ready to breed or is already paired. He picks up his spurned gift and moves on. But if the stranger receives the pebble with a deep bow then he has discovered his true mate. He bows back and the two stretch up their necks and trumpet a celebratory nuptial chorus.

One of the loveliest of the European waterbirds, the great-crested grebe, is much more elaborately costumed than the penguin. In spring, both sexes grow long chestnut-brown tippets on their cheeks, a deep brown ruff beneath the beak and a pair of hornlike tufts of glossy black feathers on the head. But again both male and female look alike. Their courtship consists of almost every manoeuvre imaginable that would show off these head adornments to good advantage. The response an individual bird gets to particular gestures tells it whether it is displaying to another bird of the same or the opposite sex. The two stretch their necks up high and twist their heads rapidly from side to side, their tippets fanned wide. They dive and pop up in front of one another. They collect strands of water plants in their beaks and present them to one another, neck stretched low over the water. And at the climax of all these ceremonies, they suddenly rear up, side by side, treading water with their feet until it looks as though they are standing on the surface, twisting their heads ecstatically from side to side.

Their courtship lasts for many weeks and elements of it are continually repeated throughout the breeding season when the birds greet one another or change places on the nest. It is as though the identically plumaged partners need to keep reassuring one another of their respective identities and the relationship between them. Even so, there are possibilities of confusion. When it comes to copulation, grebes are notorious for getting muddled, and the female may well mount the male instead of being mounted.

Close similarity of plumage is a strong indication that the birds are monogamous and that both partners share in preparing for and rearing their family. Many species, however, have some visual indicator of their sex, even if it is only a small detail like the moustache of a bearded tit, the black bib of a sparrow or the different-coloured eye of a parrot. Courtship will include displays in which the owner of this badge flaunts it in front of the partner who lacks it.

Some groups of birds have developed this sexual difference in plumage to an extraordinary degree and it is they who have brought the feather to its highest pitch of fantastication. The males of pheasants, grouse, manakins and birds of paradise grow feathers of great size and sensational colour and become so obsessed with displaying their costumes that they do little else. Their females are drab creatures who appear at the display grounds for a brief coupling and then return to lay their eggs and care for their young entirely by themselves, leaving the male still absorbed in his strutting and pirouetting, awaiting his next female visitor.

Among the most elaborate of all feathers are those grown by the male argus pheasant on his wings. Some may be over a metre long and are lined with huge eye-spots. He clears a display ground in the Borneo forest and shows off to the female by raising both wings above his head in a towering shield.

The island of New Guinea, north of Australia, contains some forty different species of birds of paradise. It is difficult to know which has the more spectacular plumage. The King of Saxony Bird, the size of a thrush, sprouts two long quills from his forehead, each bearing, on one side, a line of small enamelled blue flags; the Superb Bird has an immense emerald chest shield which it can expand until it is as broad as the bird is tall; the Twelve Wired Bird has a shimmering green bib and a huge inflatable yellow waistcoat with bare quills, the wires of its name, curling down behind it.

Watching such birds display their ornaments is one of the most thrilling and heart-stopping experiences the bird world has to offer. The New Guinea forest is, for the most part, dark and wet. Great trees soar above to cut out most of the light. But you may suddenly come across a patch of floor that has been swept clear. The leaves and litter that once covered it are piled up around the sides. It is difficult to believe that the clearing has not been made by a human being, but if you wait, the creature responsible will appear. The Magnificent Bird is the size of a starling. From his tail emerge two naked quills that curl into circles; over his shoulders he has a golden cape; on his breast, a green shield fringed with the finest shimmering blue lines. The feathers on his head and around his beak are so fine and lustrous that they look like rich black velvet. He may pause in the trees for a few minutes, hunched on a branch, assessing the situation. Then he abruptly flies to one of the saplings that grow in his court. Gripping it with both feet, he points his bill vertically upwards, spreads wide his glinting golden collar and expands his chest plumes, swelling them and contracting them so that they appear to throb, while at the same time making a buzzing noise and gaping his beak to expose the green lining of his throat. He may do this many times a day, usually in the mornings, for months on end, as will his many rivals, each with their own courts distributed through the forest, each aiming to attract females.

The most celebrated of all the birds of paradise are those with long gauzy plumes sprouting from beneath their wing coverts. There are several species, each with plumes of a different colour, yellow, red or white. These birds display communally. Their dances are held in particularly prominent trees that may have been used for the purpose for decades. One special branch in the crown will have been stripped of leaves and twigs. Soon after dawn, a flash of yellow catches your attention in the lower branches. The birds are beginning to assemble for their daily ritual. They are about the size of crows with iridescent green bibs, yellow heads and brown backs. Their golden plumes, even though they are folded, hang down on either side, doubling the length of their bodies. Soon there may be half a dozen males skulking in the undergrowth, some tentatively flicking their plumes over their backs. Eventually, one will fly up to the display branch. With a raucous shriek, he bows his head low, stropping the branch with his bill. He claps his wings above his head, throws his plumes up in a shimmering fountain of colour and scuttles up and down the branch. His passion stimulates the others to join him and soon there may be a dozen of them in the tree, shrieking and displaying, awaiting their chance to perform on the dancing branch.

A sudden movement in the shaded darkness of the branches nearby may draw your eye from this marvellous spectacle. There, plain and brown, is the hen bird. She flits across to the dancing branch and the male jumps aggressively onto her back. His plumes fall. The union lasts a second or two. Then she flies away to return to the nest that she has already prepared for her now fertilised eggs.

The male birds of paradise carry their cumbersome plumes for several months, but when the season ends, they shed them. To have to renew such large-scale accoutrements each year must make considerable demands on a bird's resources. One related group of birds in New Guinea, with similar appetites for display and polygamy, manage their affairs in what seems to be a more economical way. The bower birds achieve their results by displaying with sticks, stones, flowers, seeds and any brightly coloured objects they can find, provided that they are of a particular colour. The males construct bowers in which to display such treasures. One species stacks twigs around a sapling to form a maypole which it decorates with fragments of lichen. Another constructs a roofed grotto with two entrances in front of which he assembles flowers, mushrooms and berries, each neatly stacked in its own pile.

Other bower birds live farther south, in Australia. The male Satin Bower Bird, a dark glossy blue and about the size of a jackdaw, constructs an avenue of twigs half a metre or so apart and twice his height. He usually builds it running north and south, and at the northern, sunnier, end, he assembles his collection. There may be feathers from other birds, berries, even pieces of plastic. Their substance does not matter – only their colour. They must all be either a yellowy-green or, preferably, a shade of blue that closely matches the glint of his shining feathers. Not only does he collect such objects from far and wide and steal them from the collections of neighbours, but he sometimes mashes blue berries with his beak and uses a piece of vegetable fibre to paint the walls of his bower blue with the juice.

One way you can bring a Satin Bird down to his bower is to add to his collection an object of a quite different colour, such as a white snail shell. He usually returns very quickly and indignantly removes the aesthetically offensive object, picking it up with his bill and throwing it aside with a flick of his head. His female is, once again, a dull-looking creature. As she tours the bowers in the district, each male busies himself excitedly with his jewels, rearranging them, picking them up in his beak as though to show her their quality and calling excitedly. If he succeeds in luring her to the bower, mating takes place close by or actually between the avenue walls, accompanied by a great flapping of the male's wings, sometimes so violent that the bower walls are damaged.

The actual mechanics of mating used by birds seem clumsy. The male, with only few exceptions, has no penis. He has to mount rather precariously on the female's back, steadying himself by clinging onto her head feathers with his beak. She twists her tail to one side so that the two vents are brought together, and the sperm, with a certain amount of muscular assistance from both partners, is transferred to the female. But the process can hardly be described as neat. The female has to remain very still or the male topples off, and only too often it seems that the union is unsuccessful.

All birds lay eggs. This is the one characteristic inherited from their reptilian ancestors that no bird anywhere has abandoned. In this the birds are unique among vertebrates. Every other group has a few forms that have found it advantageous to retain eggs inside their bodies and so give birth to live young – sharks, guppies and seahorses among fish; salamanders and marsupial frogs among amphibians; skinks and rattlesnakes among the reptiles. But no bird has ever done so. Perhaps this is because a large egg inside the body, let alone a clutch of several, would be too great a load for a female to carry in flight throughout the weeks necessary for development. So as soon as the egg within her is fertilised, the female lays it.

But now the birds must pay the penalty for having developed the warm blood necessary for flight. Reptiles can bury their eggs in holes or under stones and then abandon them. Their eggs like the adults themselves, need no more than the normal heat of their surroundings to survive and develop. But the embryos of a bird have warm blood like their parents, and if they get badly chilled, they will die.

Birds therefore have to incubate their eggs, and this is a very dangerous business. It is the only time in the lives of most of them when they cannot escape their enemies by freely taking to the air. Their eggs and their young keep them sitting until the last possible moment and sometimes beyond. If they are driven to leave, their eggs and young are put at risk. Yet the nest has to be accessible so that the parents themselves can take turns in incubating and leave it to collect food for themselves and the young.

Some birds can and do nest in places that other animals find impossible to reach. Only a bird could get to a ledge in the middle of a vertical sea cliff. But there is danger even there. Some seabirds are robbers, and unless parents are careful, gulls may come and peck holes in their eggs and eat the contents.

Plovers and birds that live on sandy, gravelly shores have no alternative but to lay their eggs out in the open, for no cover exists. Their eggs are coloured to match the gravel so closely that their destruction is likely to come not from some predator that has noticed them, but from some other creature, like a blundering human, who has failed to do so and trodden on them.

Most birds, however, safeguard their eggs and young by laboriously building some kind of protection. The woodpecker excavates or enlarges holes in trees; the kingfisher bores into river banks, flying at the face with its mandibles slightly parted until it has chipped enough of a dent for it to create the foothold it needs to work with real speed. The sparrow-sized tailor bird in India sews together the growing leaves of a tree by piercing holes in their margin and tying them with separate knots of plant fibre. This forms an elegant and virtually undetectable cup within which the bird constructs its downy nest. The weaver bird, a member of the sparrow family, tears strips from palm leaves and, hanging upside down, deftly weaves them into a hollow ball, sometimes with a long vertical tube to serve as an entrance. The oven bird lives in open country in Argentina and Paraguay, where trees are few and much sought-after as homes. So it fearlessly uses fence posts and bare branches as sites and builds out of mud a near-impregnable nest, the size of a football, which resembles in miniature the oven made by local people. The entrance is large enough to admit a paw or a hand, but a cross wall inside baffles any further plunder, for the hole through it is tucked away out of line of the main entrance. Hornbills nest in holes in trees, and the male takes extreme measures to keep raiders away from the eggs and the female who is incubating them. He walls her up by building a mud wall across the entrance, leaving only a tiny hole in the centre. Through this he passes food to his long-suffering mate and nestlings. Cave swiftlets in Southeast Asia nest in caves, but since there may not be enough suitable ledges, they construct artificial ones with their glutinous spittle, sometimes mixed with a few feathers or rootlets. These are the nests that the Chinese believe make the most delectable of soups.

Some birds enlist the unwitting aid of other creatures in deterring raiders. An Australian warbler habitually builds its nests alongside those of hornets; a kingfisher in

Borneo lays its eggs actually within the nest of a particularly aggressive species of bee; and many parrots dig holes for themselves inside the brown nests of tree termites.

One family of birds has, in the most ingenious way, managed to avoid the hazardous duty of sitting on its eggs throughout the incubation period. The mallee fowl of eastern Australia lays its eggs in a large mound built by the male. The core is composed of rotting vegetation and the whole is covered with sand. The breeding season is a very long one, spread over five months, and during all this time, the male has to remain in constant attendance, probing the mound with his bill to check the temperature. In spring the newly gathered vegetation at the centre is decaying rapidly and producing so much heat that the mound may get too warm for the eggs within it, in which case he industriously removes sand from the top to allow heat to escape. In summer, there is a different danger: the sun may strike the mound and overheat it. Now he must pile more sand on top as a shield. In autumn, when the decaying core has lost much of its strength, he removes the top layers to allow the sun to warm the centre where the eggs are, and then covers it in the evening to retain the heat.

Another member of the family, living farther east on Pacific islands, has developed a specialised variation of the system. It buries its eggs in the ash on the flanks of volcanic cones and allows the lava far beneath to supply the necessary heat for its eggs.

Several species, of which the cuckoo is the most famous, have dodged the perils and labours of incubation altogether by depositing their eggs in the nest of some other bird and allowing them to rear its young. To avoid having their eggs thrown out by the foster parents they have had to develop a coloration of their eggs to match those of the species they parasitise. So each race of cuckoos restricts itself to certain species as nurses.

The process of incubation is not a straightforward one. The very fact that the bird's body feathers insulate it so well means that they form a very effective heat screen between its body and its eggs. Many therefore develop a special modification for brooding. Just before incubation starts, a group of feathers on the underside is moulted and the exposed skin becomes pink with distended blood vessels just below the surface. The eggs fit neatly into this patch and so are warmed very efficiently. But not all birds produce this patch by moulting. Ducks and geese make one mechanically by plucking their own feathers from their breasts. The blue-footed booby, which has bright blue feet and uses them in its displays, lifting them in an irresistibly comic fashion as it high-steps around its partner during courtship, now puts them to good use as incubators. It keeps its eggs warm by standing on them.

At last, the young hatch, chipping their way out of their shells with a small egg tooth on the tip of the bill. Many of those that nest on the ground are covered with down when they emerge, and this gives them excellent camouflage. They run away from their nest almost as soon as they are dry to search for food under their mother's supervision. Hatchlings of species that nest above ground in protected or inaccessible nests are often naked and helpless and have to be fed by their parents.

As the days pass, blue, blood-filled quills appear on the skins of the young and at last the essential feathers sprout. Young eagles and storks, as they fledge, may spend days standing on the edge of their nests, beating the air with their wings, strengthening their muscles and practising the movements necessary for flying. Gannets on their narrow cliff ledges do the same thing, though they prudently face inwards when they do so, just in case they become too successful too early. Such preparations, however, are the exceptions. Most young birds seem to be able to perform the complex movements of flying with virtually no practice. Some that are raised in holes, like petrels, manage to fly several kilometres at their first attempt, and almost all young birds become accomplished aeronauts within a day or so.

Astonishingly, despite their unrivalled skill in the air and all their adaptations necessary to perfect it, birds appear to abandon flight whenever possible. The older bird fossils, dating some 30 million years after Archaeopteryx but still long before the extinction of non-avian dinosaurs, included gull-like forms which were skilled flyers with a keeled chest bone and no bony tail. In essentials they were modern birds. With them, however, lived a huge swimming bird, *Hesperornis*, which was nearly as big as a man. It had already ceased flying. Fossils of those other highly successful non-flying birds, the penguins, also appear around this time.

The tendency to become grounded can still be seen in operation today. When a species of land bird colonises an island that no four-footed predators have been able to reach, sooner or later, it seems to develop into a flightless form. Rails on the islands of the Great Barrier Reef run in front of intruding feet like domestic chickens and only flutter feebly into the air on extreme provocation. The cormorants of the Galapagos have such small wings that they cannot get in the air even if they try. In New Zealand too, there were no predators before the arrival of human beings, and several bird groups there evolved flightless forms. The moas, the tallest birds that have ever existed, standing over three metres high, were hunted to extinction by the first human settlers. Only their small secretive relatives, the kiwis, survive from the whole group. There is also a strange flightless parrot, the kakapo, and a giant flightless rail, the takahe.

This relapse to a ground-living life is an indication of the great demands that flying puts on a bird's energies and the amount of food it needs in consequence. If life can be led in safety on the ground, then this is a much easier option and the birds take it. It may have been the harassing by their dinosaur relatives and by pterosaurs that drove those early feathered dinosaurs into the trees in the first place and the threat of hunting mammals that has kept their descendants there ever since.

But in between these two periods, there was an interregnum of a few million years when the dinosaurs had disappeared and the mammals had not yet developed into forms sufficiently powerful to dominate the land. It seems that the birds did then make a bid to claim the ruling position for themselves. Sixty-five million years ago, an immense flightless bird called *Diatryma* stalked the plains of Wyoming. The same creature also existed in Europe, though there it has been given a different name, *Gastornis*. It was a hunter. Taller than a man, it had a massive hatchet-shaped bill fully adequate to butcher quite large creatures.

Diatryma disappeared after a few million years, but giant flightless birds still survive elsewhere – ostriches, rheas and cassowaries. They are not close relatives of *Diatryma* but they have ancient lineages and are descended from stock that once flew. That can be deduced from the fact that they still retain many of the adaptations for flight – air sacs in the body, toothless keratin beaks and, in some instances, partially hollow bones. Their wings are not reduced forelegs but simplified versions of limbs that once beat the air, and the feathers that grow on them are arranged in the patterns appropriate for flight. The keel on their breastbone, however, has virtually disappeared, for now it has only to provide attachment for the feeblest of muscles. The feathers, since they are not needed for flight, have lost their barbules and have become merely fluffy appendages that are used in display.

The cassowaries in particular can give us some idea of how formidable a creature *Diatryma* must have been. Their feathers have lost most of their filaments and are more like coarse hair. Their stubby wings are armed with a few curving quills as thick as knitting needles. On their heads they have a bony casque with which they force their way through the thick vegetation of the New Guinea jungles where they live. The bare skin of their head and neck is livid purple, blue or yellow and hung with scarlet wattles. They feed on fruit but they also take small vertebrates such as reptiles, mammals or nestling birds. Apart from venomous snakes, they are by far the most dangerous creatures on the island. When cornered, they lash out with savage kicks that can rip open a man's stomach, and many people have been killed by them.

Cassowaries are solitary creatures. As they prowl through the forest, they often make a threatening, booming rumble which carries for considerable distances. It hardly sounds like a bird at all. Closer to, you may be able to detect the outline of an animal, as tall as a man, moving in the undergrowth. A glittering eye peers through the leaves; and then suddenly the huge creature stampedes away, crashing by brute force through bushes and saplings. You need no convincing that if large carnivorous birds developed a taste for bigger prey, they could be very dangerous animals.

Yet in the end, birds like *Diatryma* were not clever enough hunters. One group of animals escaped them. They were small insignificant creatures at that time but they were very active. Like the birds, they had developed warm blood but they insulated themselves not with feathers but with fur. They were the first mammals. It was their descendants, in the end, which inherited the earth and kept the birds, by and large, in the air.

NINE

Eggs, Pouches and Placentas

At the end of the eighteenth century, the skin of an altogether astounding animal arrived in London. It had come from the newly established colony in Australia. The creature to which it had belonged was about the size of a rabbit, with fur as thick and as fine as an otter's. Its feet were webbed and clawed; its rear vent was a single one combining both excretory and reproductive functions, a cloaca, like that of a reptile; and most outlandish of all, it had a large flat beak like a duck. It was so bizarre that some people in London dismissed it as another of those faked monsters that were confected in the Far East from bits and pieces of dissimilar creatures and then sold to gullible travellers as mermaids, sea dragons and other wonders. But careful examination of the skin showed no sign of fakery. The strange bill which seemed to fit so awkwardly on to the furry head, with a flap like a cuff at the junction, did truly belong. The animal, however improbable it might seem, was a real one.

When complete specimens became available, it was seen that the bill was not hard and bird-like as it had first seemed when the only evidence was a dried skin. In life it was pliable and leathery so the resemblance to a bird could be discounted. The fur was much more significant. Hair or fur is the hallmark of a mammal. Everyone agreed, therefore, that this mystery animal must be a member of that great group which contains creatures as diverse as shrews, lions, elephants and humans. A mammal's hairy coat insulates the body and enables it to maintain a high temperature, so it followed that this new creature must also be warm-blooded. And, presumably, it also possessed a third characteristic of a mammal and the one that gives the group its name – a mamma, a breast, with which to suckle its young.

The Australian colonists had referred to the creature as a 'water-mole' while the Aboriginal names included 'mallangong', 'tambreet' and 'dulaiwarring'. Scientists decided that they should have something that sounded a little more scholarly. There were many extraordinary features to inspire a vivid name, but the one invented for it was the rather dull one of platypus, which means no more than 'flat-footed'. Soon afterwards, it was pointed out that the name was invalid anyway as it had previously been given to a flat-footed beetle, so a second one had to be devised and the animal was re-labelled Ornithorhyncus, 'bird-bill'. This is the scientific title it still bears. To most people, however, it remains a platypus.

It lives, then as now, in the rivers of eastern Australia. Mainly nocturnal, it swims energetically and buoyantly, often cruising along the surface, paddling with its webbed forefeet and steering with its hind. When it dives, it closes its ears and tiny eyes with little muscular flaps of skin. Unable to see as it grubs around on the river bed, it feels for freshwater prawns, worms and other small creatures with its bill, which is rich in nerve endings and can detect the minute changes in pressure and the electrical signals which come from its prey. As well as being an adept swimmer it is also a powerful and industrious burrower, digging extensive tunnels through the river banks sometimes as much as 18 metres long. To do this, it rolls back the webbing of its forefeet into its palms and so frees the claws for work. At the end of the tunnel, the female constructs an underground nest of grass and reeds. From one of these came more sensational news about the animal. It was claimed that the platypus laid eggs.

Many zoologists in Europe regarded this as being altogether too absurd. No mammal laid eggs. If eggs were found in a platypus nest, then they must have been deposited there by some other visiting creature. They were described as being nearly spherical, about the size of marbles and soft-shelled, in which case they were probably those of a reptile. But local people in Australia insisted that they belonged to the platypus. Naturalists argued heatedly about the issue for nearly a century. Then in 1884, a female was shot just after she had laid an egg. A second one was found inside her body on the point of being extruded. Now there could be no doubt. Here was a mammal that did indeed lay eggs.

Further surprises were to come. When, after ten days, these eggs hatch, the young are not left to find food for themselves as all young reptiles must do. Instead, the adult female develops some special glands on her belly. They are similar in structure to the sweat glands that the platypus, like most mammals, has in its skin to aid in cooling the body if it gets overheated. But the sweat these enlarged glands produce is thick and rich in fat; it is milk. It oozes into the fur and the young suck it from tufts of hair. There is no nipple, so the platypus cannot be said to have a true breast; but it is a beginning.

That other important mammalian character, endo-thermy or warm-bloodedness, also seems to be incompletely developed. Nearly all mammals keep their bodies at temper-atures between 36° and 39°C. The platypus' temperature is only 30°C and fluctuates very considerably.

One other creature in the world can parallel this mixture of primitive mammalian and reptilian features, and it, too, comes from Australia: the spiny anteater. The history of its naming is a repeat of that of the platypus. Science first called it echidna, 'spiny one', only to discover that name had been bestowed previously on a fish. So it was renamed tachyglossus, 'swift-tongued'. But once more it was the first name that stuck. The animal looks like a large flattened hedgehog with an armoury of spines on its back embedded in a coat of dark bristly hair. It can dig itself into the ground with swimming movements of its four legs, which excavate so efficiently and with such strength that on anything but the hardest surface, an alarmed echidna simply goes down vertically and within a few minutes all that can be seen of it is an impregnable dome of extremely sharp spines.

The animal is not primarily a burrower. It goes to ground largely as a defensive measure. Most of its time it spends either asleep in some unobtrusive corner or waddling through the bush searching for ants and termites. When it finds a nest of them, it tears it open with the claws of its front legs and licks up the insects with a long tongue which flickers in and out of the tiny mouth at the end of its tube-like snout. This snout and its spines, like the bill of the platypus, are specialised characteristics that befit it for its particular way of life. In evolutionary terms, they are recent acquisitions. Fundamentally, the echidna is very similar to the platypus. It has hair; its body temperature is very low; it has a single vent, the cloaca; and it lays eggs.

In one detail of its reproduction, it differs. The female lays a single egg which she keeps not in a nest but in a temporary pouch which develops on her underside. When the moment for laying arrives, she curls round and manages to deposit the egg directly into her abdominal pouch, thus exhibiting a gymnastic ability one might not have suspected in such a comfortably plump creature. The shell of the egg is moist and sticks to the hair in the pouch. After seven to ten days, it hatches. Thick yellowish milk exudes from the skin of the mother's belly and the hatchling sucks it up. It remains in the pouch for some seven weeks, by which time it is about ten centimetres long and its spines have begun to develop. This presumably makes the puggle, as local people call it, an uncomfortable passenger as far as its mother is concerned. At any rate, the mother now removes it and deposits it in a den while she goes out to feed. She returns about once a week to feed it, and continues to do so for several more weeks, encouraging it to suckle by prodding it beneath her body with her snout and arching her back so that her abdomen is clear of the ground. The puggle then lifts its head and fastens its little jaws on tufts of her hair.

In contrast, the only food a reptile provides for its baby is the yolk in the egg. From this small yellow ball, the young reptile must build a body that is sufficiently complete and strong to make it totally independent as soon as it emerges from the shell. It must then go and seek food for itself – nearly always of the same kind as it will live on for the rest of its life. The platypus' method has much greater potential. Its eggs have a small amount of yolk in them, but by providing its young as soon as they hatch with a continuous supply of special easily digested food, the milk, it allows them to have a much longer development. This is a major change in maternal technique and one that has been crucial, in even more sophisticated versions, to the ultimate success of the whole mammal group.

The body design of the echidna and platypus is undoubtedly of great antiquity, but we have no hard evidence to indicate which fossil reptiles were their ancestors. Our knowledge of many of the candidates is based to a considerable degree on teeth. These are among the most durable parts of any animal's anatomy, and therefore frequently preserved as fossils. They are very informative about an animal's diet and habits, and also highly characteristic of a species and strong evidence of genealogical relationships. Inconveniently, when the platypus and echidna became specialised – for under-water foraging in one case and ant-eating in the other – they both lost their teeth. Their ancestors assuredly once possessed them, for young platypus still produce three tiny ones soon after birth, but they are lost within a very short time and replaced with horny plates. Nor is there fossil evidence of any consequence about these ancestors. So we have virtually nothing to help us link these creatures to any particular group of fossil reptiles. Nonetheless, it is a reasonable guess that the kind of breeding techniques the platypus and echidna use today were developed by some reptile groups in the process of their transformation into mammals.

But which reptiles were they? The hallmarks of today's mammals – such as hair and milk-producing glands – do not fossilise. But we can trace the origins of the genes that produce them. These reveal that the mammals and the reptiles split over 300 million years ago as different forms of the gene that produces keratin appeared. In reptiles and birds, keratin makes feathers and scales. In mammals, a slightly different form of it makes hair. At the same time, mammals evolved glands at the roots of their hair that eventually gave rise to sweat and milk glands.

Finding the origin of that other mammalian charac-teristic – endothermy or warm-bloodedness – is more complicated, for there is no single gene or group of genes involved and there are no fossilised structures. However, one group of reptiles from which the mammals developed may have been able to do this. These were the pelycosaurs. One of them, dimetrodon, grew long upright spines from its backbone which supported a sail of skin. This may have served as a solar panel, absorbing heat, but this has yet to be proved. However, not all pelycosaurs had sails. It is therefore supposed that the pelycosaurs and their successors, the therapsids, were to some degree endothermic. The therapsids were only a metre long. Endothermy, particularly in such small creatures, needs some form of bodily insulation if it is to be effective, so it could be that some of these creatures were covered in hair. The genes for producing it had certainly evolved by this time. We can only deduce their presence.

There are other clues to suggest that some of the therapsids were on the way to becoming mammals. Generating heat in the body, in a proper endothermic manner, absorbs a lot of energy and would have required both an increase in daily food and a speeding up of the digestive processes. One way of achieving this would be to replace the typical reptilian teeth, which are simple and peg-like and do little more than grip, with specialised cutters, grinders and mashers that can break up the food mechanically. And this is exactly the change that can be traced in the teeth of therapsids.

But even supposing that they were both warm-blooded and hairy, would that make them mammals? The question, of course, is to some extent an artificial one. These categories are our inventions, not nature's. In practice, ancestral lines merge imperceptibly into one another. The anatomical characters which, grouped together, we may choose to regard as diagnostic, may individually change at different speeds so that one feature may develop while the remainder of an animal's anatomy remains relatively unmodified. Furthermore, the environmental conditions that stimulate such change may produce similar responses in several dynasties. Indeed, there seems no doubt that warm blood was acquired at various times by several quite separate reptilian groups. It may therefore be that the line of reptiles from which the platypus and echidna stemmed was not the same as that which was to give rise to other mammals.

By 200 million years ago, fully formed mammals had appeared. A small fossil discovered in 2013 in China and dating from 160 million years ago, is the earliest near-complete specimen of a mammal that has so far been found. This creature, named rugosodon ('wrinkle-tooth'), was only about 17 centimetres long and somewhat shrew-like. Its teeth suggest that it was an omnivore that could eat plants as well as insects and worms and it must have been both warm-blooded and furry. It was not an ancestor of any living mammal but belonged to a group known as the multituberculates, which survived for over 160 million years before becoming extinct around 30 million years ago. Whatever its eventual fate, its presence shows that the mammals had arrived.

Even so, the next great developments among the land animals did not come from them. Instead, the dinosaurs and the other reptiles – the flying pterosaurs and the marine reptiles – began their dramatic expansion. Although the mammals were quite overshadowed in both numbers and size, they survived, saved by their warm blood, which allowed them to be active at night, when the great reptiles may have been torpid. These small warm-blooded creatures may then have emerged from hiding to hunt for insects and other small creatures. Some mammals, such as one called repenomamus, were the size of a cat and may even have eaten small dinosaurs. But essentially, these early mammals lived in the shadow of the reptiles.

This situation prevailed for a vast period of time – 135 million years – until the cataclysm that brought an end to the non-avian dinosaurs and so much else. The little mammals, however, survived it and quickly evolved into new forms to reoccupy vacated places in the world's ecosystems.

Among them were creatures that were very like the opossums that today live in the Americas. The Virginia opossum is a large, rat-shaped creature, much bewhiskered, with an untidy shaggy coat, button eyes and a long naked tail which it can wrap round a branch with sufficient strength to support its own weight – for a little time at least. It has a large mouth that it opens alarmingly wide to expose a great number of small sharp teeth. It is a tough, adaptable creature that has spread through the Americas, from Argentina in the south to Canada in the north. There it endures such cold temperatures that on occasion its large,

Above: *Hourglass frog calling with its throat-sac inflated to amplify its calls, Costa Rica. (Alex Hyde/ naturepl.com)*

Right: *Male three-striped arrow-poison frog with his tadpoles on his back, Amazon rainforest, Peru. (Konrad Wothe/ naturepl.com)*

Previous page: *Poison-arrow frog, Costa Rica. (Visuals Unlimited/ naturepl.com)*

Left: *Marine iguana renewing its energy by basking in the sun on Fernandina Island, Galapagos Islands.* (Franco Banfi/naturepl.com)

Below: *David Attenborough with a leatherback sea turtle.* (Gavin Thurston)

Opposite: *Nile crocodile launching itself into a river, Tanzania. (Charlie Summers/naturepl.com)*

Above: *Panther chameleon, Madagascar. (Alex Hyde/naturepl.com)*

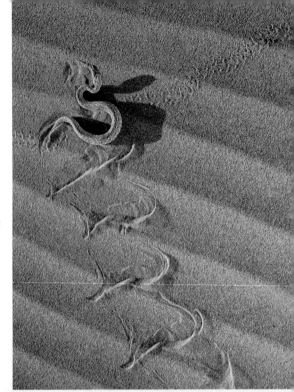

Right: *Side-winding snake travelling across a sand dune, Namibia. (Emanuele Biggi/naturepl.com)*

Below: *Underside view of a gecko's toes, clinging to glass. (Stephen Dalton/naturepl.com)*

Opposite: *Cape cobra displaying its hood, South Africa. (Tony Phelps/naturepl.com)*

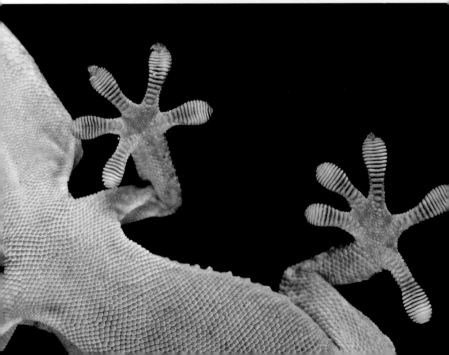

Opposite: *The bones of a 150 million-year-old bird-like dinosaur from the late Jurassic, with impressions of feathers around its limbs and body, Liaoning Province, China. (Martin Shields/Science Photo Library)*

Left: *Hoatzin chick, clambering through the mangrove branches around its nest, Guyana. (Flip de Nooyer/Minden Pictures/FLPA)*

Below: *Common cranes migrating across the Pyrenees, France. (Nick Upton/naturepl.com)*

Skimmer slicing through the surface of a river in search of fish, Pantanal, Brazil.
(Tony Heald/naturepl.com)

Sword-billed hummingbird, about to collect nectar from the flower of a Datura plant, Ecuador. (Nick Garbutt/naturepl.com)

Great-crested grebes presenting weed during their courtship ritual, Derbyshire, UK. (Andy Parkinson/2020 VISION/naturepl.com)

Left: *The courtship display of a male argus pheasant, Danum Valley, Borneo. (Juan Carlos Munoz/naturepl.com)*

Below: *A greater sage-grouse in full display, Wyoming, USA. (Gerrit Vyn/naturepl.com)*

Above: *Male Red Bird-of-paradise displaying his plumes. (Tim Laman)*

Right: *The four-egg clutch of a little ringed plover on shingle, Lorraine, France. (Michel Poinsignon/naturepl.com)*

Left: *Greater flamingos mating, Camargue, France. (Theo Allofs/naturepl.com)*

Below: *Male kakapo, Codfish Island, New Zealand. (Brent Stephenson/ naturepl.com)*

Opposite: *Male baya weaver, weaving his nest, India. (Ingo Arndt/naturepl.com)*

Above: *Short-beaked echidna, starting to dig a hole, Tasmania, Australia. (Dave Watts/naturepl.com)*

Left: *Dimetrodon, a pelycosaur dinosaur, 290 to 248 million years old. (John Weinstein/Field Museum Library/Getty Images)*

Above: *Female Virginia opossum carrying her young on her back, Minnesota, USA. (ARCO/naturepl.com)*

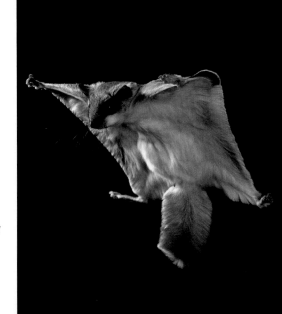

Right: *Northern flying squirrel in flight, Canada. (Stephen Dalton/naturepl.com)*

Above: *Young western grey kangaroo inspects his mother's pouch that has been his nursery, Kangaroo Island, South Australia. (Aflo/naturepl.com)*

Left: *Numbat searching for food, central Australia. (Jouan Rius/naturepl.com)*

Above: *Tupaia, Borneo. (Rod Williams/naturepl.com)*

Below: *Yellow-streaked tenrec eating a worm, Madagascar. (Pete Oxford/naturepl.com)*

Opposite: *Pangolin, India. (Yashpal Rathore/naturepl.com)*

Above: *Giant anteater, Brazil. (Nick Garbutt/naturepl.com)*

Right: *Greater horseshoe bat in flight, catching a moth, Germany. (Dietmar Nill/ naturepl.com)*

Opposite top: *Straw-coloured fruit bats returning to their daytime roost at sunrise, Zambia. (Nick Garbutt/ naturepl.com)*

Opposite below: *Pod of humpback whales feeding communally on herring, Chatham Strait, Alaska, USA. (Tony Wu/naturepl.com)*

Above: *Family of sperm whales, Indian Ocean. (Tony Wu/naturepl.com)*

Below: *Three-toed sloth, the fur on its back stained faintly green by algae, Costa Rica. (Suzi Eszterhas/naturepl.com)*

Jaguar in the rainforest of Central America. (Tom & Pat Leeson/Science Photo Library)

Sumatran rhinoceros, Kambas National Park, Sumatra, Indonesia. (Cyril Ruoso/naturepl.com)

Bengal tiger, Ranthambore National Park, India. (Andy Rouse/naturepl.com)

Above: *Leopard catching a springbok, Etosha, Namibia. (Wim van den Heever/naturepl.com)*

Below: *Lionesses stalking a herd of zebra, Masai Mara, Kenya. (Frans Lanting, Mint Images/ Science Photo Library)*

Above: *Sifaka jumping feet foremost into a tree, northwest Madagascar. (David Pattyn/ naturepl.com)*

Left: *Tarsier, Bohol Island, Philippines. (David Tipling/ naturepl.com)*

Opposite: *Golden-headed lion tamarin, Brazil. (Juan Carlos Munoz/ naturepl.com)*

Left: *Mother orang utan helping her baby to climb, Guning Leuser, Sumatra. (Cyril Ruoso/naturepl.com)*

Below: *Bald-headed uakari monkey, Brazil. (Roland Seitre/naturepl.com)*

Above: *Mother chimpanzee catching termites with a grass stem while her youngster watches and learns, Gombe National Park, Tanzania. (Anup Shah/naturepl.com)*

Below: *Arabian oryx, Abu Dhabi, United Arab Emirates. (Fabian von Poser/Imagebroker/FLPA)*

naked ears get frost-bitten. It wanders around the country-side with a raffish buccaneering air, eating fruit, insects, worms, frogs, lizards, young birds – almost anything that might be considered edible.

The most extraordinary thing about it, however, is the way it reproduces. The female has a capacious pouch on her underside in which she rears her young. When the first opossum was brought to Europe from Brazil in the early sixteenth century by the explorer Martin Pinzón, who had served under Columbus, no one had ever seen anything like it. The king and queen of Spain were persuaded to put their fingers inside the pouch and wonder. Scholars gave the structure a name, marsupium, 'little bag', and so the opossum became the first marsupial to be known in Europe.

There was no doubt that the young were reared in the marsupium, for they were often found there, minuscule naked pink creatures clinging to the teats with their mouths; but how did they get there? Some said at the time, and some country folk in America still maintain, that they are, literally, blown into the pouch. The opossums mate, according to the story, by rubbing noses. The young are conceived in the nostrils and in due course the female sticks her nose in her pouch, gives a hefty snort and blows out the lot of them. The story doubtless arose because the female opossum, just before the young appear in the pouch, pokes her nose into it and carefully licks it clean to prepare for their arrival.

The truth is scarcely less fantastic than the fable. Opossums, like the echidna and platypus, have a single cloaca closed by a sphincter muscle into which the anus and the urino-genital vent both open. They copulate and the male fertilises the female's eggs internally, but the young embryos that result have only tiny yolk sacs to supply them and they are expelled into the world after twelve days and eighteen hours, the shortest gestation period known in any mammal. Blind pink morsels no bigger than bees, they are so unformed that they cannot justly be called infants or babies, and are referred to instead by the special name of neonate. The female may produce as many as two dozen of them at a time. As they emerge from their mother's cloaca, they haul themselves through the fur of her belly to the opening of the pouch, a distance of about eight centimetres. It is the first and the most hazardous journey of their lives and half of them may well die on the way. Once they reach the warmth and security of the pouch, each fastens on to one of the mother's nipples and starts to take milk. Their mother has only thirteen nipples, twelve forming a circle with the thirteenth in the middle. If more than thirteen complete the journey, the latecomers, finding no vacant teat, will starve and die.

Nine or ten weeks later, the young clamber out of the pouch. They are now fully formed, the size of mice, and cling to their mother's fur in what seems to be a most precarious fashion. In the early eighteenth century, a famous illustration of a South American opossum showed the young with their tiny tails neatly wrapped around their mother's tail which trails behind her. As this was copied by one illustrator after another, this posture became transformed into one in which the mother arches her tail over her back and the babies in a neat row strap-hang from it by theirs. When museums came to mount opossum skins, they consulted books and understandably mounted their specimens in this engaging posture, so giving further strength to the story. It is, however, only another of the fables with which this odd creature seems to be surrounded. Young opossums are not nearly so orderly. They clamber all over their mother, clinging to her long fur, sometimes beneath her, sometimes on her back, with as much abandon and contempt for normal safety as children cavorting in an adventure playground. It is three months before they leave her and set off on an independent life of their own.

There are over a hundred different species of opossum in the Americas. The smallest is mouse-sized and does not have a pouch. Its young, no bigger than grains of rice, cling to the teats between their mother's hindlegs and hang there like a diminutive bunch of grapes. At the other end of the scale, the water opossum is almost the size of a small otter. It has webbed feet and spends much of its time swimming. Its young are saved from drowning by one of the more elaborate of pouches. It is closed by a sphincter, a ring-shaped muscle which shuts the entrance like the drawstrings of a purse. The young inside are able to endure several minutes of submergence and breathe air containing a concentration of carbon dioxide that would stifle most creatures.

The earliest mammalian fossils that have been certainly identified as being marsupial were found in South America, and it appears that the group originated here.

But the greatest assemblage of marsupials today live not in America but in Australia. How could they have got from one continent to the other?

To find an answer to that question, we have to return to the period when the dinosaurs were still at the height of their dominance. This was the time when the continents of the world were in contact with one another and formed a supercontinent which geologists call Pangaea. So it is that fossils of closely related dinosaurs have been found in all of today's continents, in North America as well as Australia, in Europe as well as Africa. The reptilian ancestors of the mammals must have been similarly widespread. But towards the end of the dinosaurs' reign, this great land split into two. The northern half, called Laurasia, contained today's Europe, Asia and North America; the southern, named Gondwana, would eventually fragment into South America, Africa, Antarctica and Australia.

The primary evidence for this grouping, and the subsequent splitting and drifting of the continents we know today, is geological. It comes from studies of the way in which today's continents fit together, the continuities of the rocks between their opposite edges, the orientation of magnetic crystals in rocks which show the position they held on the globe when they first formed, the dating of the mid-ocean ridges and the islands that rise from them, drillings in the ocean floors and other sources.

Animals and plants also add corroborative evidence. Giant flightless birds provide a particularly clear case. As we have seen, they appeared very early in the history of the birds. One group which included the ferocious diatryma evolved in the northern supercontinent. All of these are now extinct. In the southern supercontinent, Gondwana, another different family appeared which fared much better. These were the ratites. This group includes the rhea in South America, the ostrich in Africa, the emu and cassowary in Australia and the kiwi in New Zealand. Although these flightless birds were once suspected to have simply walked to their current homes when the continents were joined, it is now thought that their flightlessness is an example of convergent evolution – their ancestors flew to these locations and only then did each group lose its ability to fly.

Fleas, too, support the case. These parasitic flightless insects travel with the animals they live on but they also readily transfer to new hosts and develop into new species if they get the chance. Some families of highly characteristic fleas are found only in Australia and South America. But nowhere in between. If their hosts had carried them through Europe and North America, the only alternative route between the two continents, they would surely have left relatives among other furred creatures on the way. But there are none.

Botany also provides evidence. The southern beeches are a group of trees and shrubs related to but very distinct from the European beech. Some are found in South Africa, others in New Zealand and Australia. But none elsewhere. The same is true of those semi-tropical plants the proteas and banksias, whose spectacular flowers are so strikingly similar. The first are South African, the second Australian.

Gondwana eventually broke up still further. Africa separated and drifted northwards. Australia and Antarctica remained joined to one another and were linked, either by a land bridge or a chain of islands, to the southern tip of South America. At this point, it seems, the marsupials were still developing from the early mammal stock. Some evidence suggests that this happened in the part of Gondwana that would become South America and that they then spread into the other parts that would eventually become Australia and Antarctica.

Meanwhile, primitive mammals were also evolving in the northern supercontinent. They were to develop a different way of nourishing their young. Instead of transferring them at a very early stage into an external pouch, they retained them within the body of the female and supported them by means of a device called the placenta, a technique we will examine later.

The South American marsupials flourished greatly while they had no competition from any other kind of mammal. A huge wolf-like form appeared, and also a leopard-like carnivore with sabre-like canine teeth. But the fragments of the southern supercontinent were drifting apart and South America was moving slowly northwards. In due course, it connected with North America by way of a land bridge in the neighbourhood of Panama. Down this corridor came the placental mammals to dispute the possession of South America with the marsupial residents. In the course of this rivalry, many species of marsupials disappeared, leaving only the tough, opportunist opossums. Some of these invaded the land of the invaders and managed to colonise North America, as the Virginia opossum has done today.

The marsupials that lived in the central part of the southern supercontinent, however, did not survive at all. This great block of land became Antarctica. It drifted over the South Pole where it was so cold that it developed an immense icecap and life on the land became insupportable. But geologists have now found fossils to show that this was once marsupial territory.

The creatures on the third section of the supercontinent, however, were more fortunate. This was Australia. It drifted north and east into the emptiness of the Pacific basin and remained totally separate from any other continent. So its marsupials have evolved in isolation for the last 50 million years. During this vast period, they developed into a great number of different types, taking advantage of the wide range of environments available to them. You can see the remains of some of the spectacular species that once existed in the limestone caves of Naracoorte, 250 kilometres south of Adelaide. These caves have been famous for years because of the beauty of their stalactite formations, but in 1969, a faint breath of air, filtering up through boulders at the far end of the main chamber, hinted that there might be hitherto unknown sections beyond.

Excavations revealed a narrow passageway which eventually led to the greatest assemblage of marsupial fossils yet found. After an hour of crawling on hands and knees, wriggling through tight rock squeezes and worming down narrow winding chimneys, you come at last to two low galleries. You can only enter them by inching through a narrow tunnel on your stomach. Before you stretches a long gallery not much more than a metre high with a ceiling hung with straw-like stalactites. The air is so humid that your breath turns to mist in front of you and a party of half a dozen human cavers could, in a few minutes, cause the entire gallery to fill with fog. The floor is covered with a soft red silt, carried down here by floods of a subterranean river that has long since disappeared. With the mud, it brought the bones of marsupials. Some were animals that had lived in the upper cave. Others appear to have been creatures of the surrounding forests that had accidentally fallen down swallow holes at the cave's entrance and been killed. The bones lie strewn thickly on the mud – leg bones, shoulder blades, teeth and, most dramatic of all, skulls. All are a delicate pale cream colour as though they were speci-mens straight from an anatomist's cleaning bath. Most are so fragile that they crumble at a touch and can only be safely lifted if they are jacketed in foam and plaster.

There are the remains of a huge marsupial, the size and shape of a cow, of an immense kangaroo with a head like a small giraffe that browsed on the branches of trees. The character of one creature here is still under discussion. It was originally thought to be a carnivore, for its back teeth are elongated into formidable shearing blades with which it might have sliced through the flesh and bones of its prey. Because of its size, it was called a marsupial lion. Now studies of its front legs have shown that they were well suited for clinging, so it may actually have been a tree climber and used its fearsome-looking back teeth merely to cut hard fruits.

These creatures died about 40,000 years ago. The factors that ultimately brought about their extinction are still uncertain. It may be that the animals were affected by changes in the climate. Australia, after it had broken away from Antarctica, continued to drift northwards. Indeed, it is still doing so and as fast as it has ever done, at a rate of about five centimetres a year. The shift brought about a gradual warming and drying of the continent. It may also be that these extinctions were hastened by human beings when they first arrived there, 40,000 or more years ago.

Great numbers of marsupials still survive, of course. Today there are a dozen main families with, between them, nearly two hundred species. Many of these creatures parallel the placental forms that evolved in the northern hemisphere. When colonists from Europe came to Australia, they often ignored the names used by the Aboriginal people and referred to the marsupials by the names of the European creatures they most closely resembled. In the temperate forest of the south, for instance, the colonists found little furry, pointed-nosed long-tailed creatures and called them, understandably, marsupial mice. The name is not really appropriate, for these are not rodents, timidly nibbling grain, but savage hunters that will set upon insects quite as large as themselves and crunch them to pieces. There are also carnivorous marsupials that will tackle reptiles and nestling birds and are called marsupial cats.

In two instances, the parallels between placental and marsupial forms are so close that if you came across one of them in a zoo you would find it almost impossible to tell which it was without handling it. The sugar glider is a small leaf- and blossom-eating marsupial that lives in eucalyptus trees. It has a parachute of skin connecting its fore- and hindlegs which enables it to glide from branch to branch. It looks almost exactly like a North American flying squirrel. A burrowing way of life demands particular structures, and both marsupial and placental burrowers have developed them. Moles of both kinds have short silky fur, reduced eyes, powerful digging forelegs and a stumpy tail. The female marsupial mole, however, has a pouch and one which, fortunately for her young, opens backwards so that it does not fill with earth as she burrows through her tunnels.

Not all marsupials have such close placental equivalents. The koala is a medium-sized creature that lives in trees and feeds on leaves, a role that is filled elsewhere by monkeys. But the koala is scarcely monkey-like in appearance and its slow-moving plodding character is a long way from that of the intelligent and quick-reacting monkeys. The numbat is an anteater. It has the long sticky tongue that all ant-eaters use to collect their food, but its adaptations are not nearly so extreme as those of, for example, the giant anteater of South America which has developed a long curving tube for a snout and lost all its teeth. The numbat's jaws are not nearly so elongated and it still has all its teeth. One marsupial, the honey possum, has no placental equivalent at all. It is only the size of a mouse, its jaws are pointed and it has a tongue with a brush on the end of it, like that of some parakeets, with which it licks nectar and pollen from flowers.

In the temperate woodlands of Tasmania lives another creature which is also uniquely and quintessentially Australasian, the boodie. It is one of a small group of marsupials called, collectively, rat kangaroos. Very shy, strictly nocturnal, it feeds on all kinds of foods, including meat, and it has a pair of small pointed canine teeth to help it do so. It makes its nest in a burrow, industriously collecting material for it in a most ingenious way. It picks up a few straws in its mouth, stacks them in a bundle on the ground and then pushes them back over its long tail with its hindlegs. The tail then curls up tightly so that the straw is effectively baled and the boodie moves away. To do so, it hops. Boodies move about entirely on their back legs, which have very long feet. If you had to design a creature to serve as a primitive ancestor of the most famous Australian animal of all, the kangaroo, it might well look like the shy, forest-living, omnivorous, hopping boodie.

The development of the kangaroo clan was accelerated by Australia's continuing drift northwards and the consequent drying and warming of its climate. This caused the forest that had covered much of the land to thin and be replaced by more open country and grassland. Grass is good food, but to move out of the forest and graze in the open is to be exposed to attack by hunting animals. So any grass-eaters that colonised the plains would have to be able to move fast. The kangaroos achieved that with a greatly exaggerated version of the boodie method. They hopped – and prodigiously.

No one really knows why kangaroos use this method rather than running on all fours, as virtually all herbivorous plain-living creatures do elsewhere in the world. The European colonists could find no equivalent in their bestiaries so they had to adopt one of the names the Aboriginal people used – gangurru. Maybe the tendency to an upright stance was already there in their ancestors, as it is in the boodie, though such a thought only puts the question back a stage. Maybe hopping is connected with the problems of carrying large babies in a pouch, which might be more conveniently done, particularly when moving at high speed over rough and rocky ground, with an upright torso. Whatever the reason, the kangaroos have brought the hop to a high pitch of efficiency. Their hindlegs are enormously powerful, the long muscular tail is held out stiffly behind so that it acts as a counterbalance, and the animal, in bursts, can reach speeds of 60 kph and clear fences nearly 3 metres high.

The second difficulty that grass-eaters must overcome is the wear and tear on their teeth. Grass is tough, particularly the kind that grows today in the parched land of central Australia. Breaking it down into pulp in the mouth is a very valuable aid to its digestion, but it is very wearing on the teeth. Grazers elsewhere have molars with open roots so that wear can be compensated by continuous growth throughout the animal's life. Kangaroo teeth have no such ability. Their roots are closed, so they use a different system of replacement. There are four pairs of cheek teeth on either side of the jaws. Only the front ones engage. As they are worn down to the roots, they fall out and those from the rear migrate forward to take their place. By the time the animal is fifteen or twenty years old, its last molars are in use. Eventually these too will be worn down and shed, so that even if the venerable animal does not die for any other reason, it will do so from starvation.

There are some forty different species in the kangaroo family. The smaller ones are usually called wallabies. The largest is the red kangaroo, which stands taller than a man and is the biggest of all living marsupials.

Kangaroos reproduce in much the same way as the opossums. The egg, which is still enclosed in the vestiges of a shell a few microns thick and has a small quantity of yolk within it, descends from the ovary into the uterus. There, lying free, it is fertilised and begins its development. If this is the first time that the female has mated, it does not stay there long. In the case of the red kangaroo it is only thirty-three days before the neonate emerges. Usually only one is born at a time. It is a blind, naked worm a few centimetres long. Its hindlegs are mere buds. Its forelegs are better developed and with these it hauls its way through the thick fur of its mother's abdomen. She herself seems entirely oblivious to it. It was once thought that she at least helped it on its way by licking a pathway through her fur, but now it is known that when she licks her abdomen, she is only cleaning herself up after the release of fluids from the ruptured egg membranes that have oozed from her cloaca.

The neonate's journey to the pouch takes about three minutes. Once there, it fastens on to one of four teats and starts to feed. Almost immediately, the mother's sexual cycle starts again. Another egg descends into the uterus and she becomes sexually receptive; she mates and the egg is fertilised. But then an extraordinary thing happens. The egg's development stops.

Meanwhile, the neonate in the pouch is growing prodigiously. The teat is a long one and has a slight swelling on the end so that, if it is pulled out carelessly, the neonate's mouth may be torn and bleed slightly. But there is no truth in the story that mother and offspring become fused together, or that milk is pumped into the young under pressure.

After 190 days, the baby is sufficiently large and independent to make its first foray out of the pouch. From then on it spends increasing time in the outside world and eventually, after 235 days, it leaves the pouch for the last time.

If there is a drought at this time, as happens so often in central Australia, the fertilised egg in the uterus still remains dormant. But if there has been rain and there is good pasture, then the egg now restarts its development. Thirty-three days later, another bean-sized neonate will wriggle out of the mother's cloaca and make its laborious and risky way up to the pouch. The female will then immediately mate again. But the first-born does not give up its milk supply so easily. It returns regularly to feed from its own teat. What is more, the milk with which it is now supplied is a different mixture from that which it first received. So now the female has three young dependent on her: one active young-at-foot which grazes but comes back to suckle; a second, the tiny neonate, sucking at her teat in the pouch; and a third, the fertilised but undeveloped egg, waiting its moment within her uterus.

It is a commonly held notion that the marsupials are backward creatures, scarcely much of an improvement on those primitive egg-layers, the platypus and echidna. This is a long way from the truth. The marsupial method of reproduction must certainly have appeared very early in mammalian history, but the kangaroos have refined it marvellously. No other creature anywhere can compare with the female kangaroo which, for much of her adult life, supports a family of three in varying stages of development.

The mammalian body is a very complicated machine that takes a long time to develop. Even as an embryo it is warm-blooded and burns up fuel very quickly. Both these characteristics demand that the developing young should be supplied with considerable quantities of food. All mammals have found methods of providing far more than could ever be packed within the confines of a shelled egg. The early mammals in the northern supercontinent included marsupials. However it is extremely unlikely that they reached such sophisticated levels of efficiency as those achieved by the Australian marsupials today.

The method developed in the north, however, allows the young to remain in the uterus for a very long time. It does so by means of a placenta, a flat disc that becomes attached to the wall of the uterus and connected by the umbilical cord to the foetus. The junction with the uterine wall is highly convoluted so that the surface area between the placenta and the maternal tissues is very great. It is here that interchange between mother and foetus takes place. Blood itself does not pass from mother to young, but oxygen from her lungs and nutrients derived from her food, both dissolved in her blood, diffuse across the junction and so enter the blood of the foetus. There is also traffic in the other direction. The waste products produced by the foetus are absorbed by the mother's blood and then excreted through her kidneys.

All this makes for great biochemical complications. But there are further ones. The mammalian sexual cycle involves the regular production of a new egg. This causes no problem to the marsupial, for in every species, the neonate emerges before the next egg is due to be produced. The placental foetus, however, stays in the uterus for very much longer. So the placenta secretes a hormone which suspends the mother's sexual cycle for as long as the placenta is in place so that no more eggs are produced to compete with the foetus in the uterus.

There is also another problem. The foetus' tissues are not the same, genetically, as the mother's. They contain elements from the father. So when it becomes connected to the mother's body, it risks immunological rejection in the same way as a transplant does. Just how the placenta prevents this from happening is still not fully understood, but it seems that a specific part of the mother's immune response is turned off at the very beginning of pregnancy.

So, by these means, the babies of placental mammals can remain in the uterus until, if necessary, they are so well developed that they can be fully mobile as soon as they are born. Even after this, they are provided with milk for a further period until they have to gather food for themselves from the world around them.

The placental breeding technique spares the young the hazardous journey outside their mother's body at a very early stage that a marsupial neonate has to undertake, and allows their mother to supply their every want during the long period they remain within her. So whales and seals can carry their unborn young even as they swim for months through freezing seas. No marsupial with air-breathing neonates in a pouch could ever succeed in doing such a thing. It was this placental technique that, in the end, was to prove one of the crucial factors in the mammals' ultimate success in colonising the whole of the earth.

TEN

Theme and Variation

S it quiet and motionless in a forest in Borneo and you have a fair chance of being visited by a small, furry, long-tailed creature which runs four-footedly along the branches of the bushes and over the ground, inquisitively testing everything with its pointed nose. It looks and behaves rather like a squirrel. A sudden unexpected noise makes it freeze, its glittering button-sized eyes wide with alarm. Equally suddenly it jerks back into frantic activity, flicking its tail backwards and forwards as it moves. But if when it finds something to eat, it does not nibble at it with its front teeth but opens its mouth wide and champs vigorously with huge relish, then you are watching something much more unusual than a squirrel and a creature of some significance – a tupaia.

If ever there were a creature that has been all things to all men, this is it. Local people in Borneo very understandably, regard it as a kind of squirrel; they use the word *tupai*, which science has adopted, for all such animals. The first European scientists to catch a specimen, discovering that it lacked the gnawing teeth of a rodent and had numerous small spiky ones, called it a tree shrew. Other people believed that some details of its genitals indicated a relationship with marsupials. Nearly a century ago, a very eminent anatomist analysed the structure of its skull in great detail, noted that the creature had a surprisingly large brain, and argued that it should be regarded as an ancestor of monkeys and apes and classified it with them.

In recent years, molecular analysis has helped to swing the balance of opinion away from viewing it as an ancestral monkey and regards it instead as a sister group of rodents and rabbits. Nevertheless, the first mammals that scampered about in the dinosaur-dominated forests may have looked very like it – small, long-tailed and pointed-nosed, and, by inference, furry, warm-blooded, active and insect-eating.

The reign of the reptiles had been a long one. They had come to power about 250 million years ago. They had browsed the forests and munched the lush vegetation of the swamps. Meat-eating forms had developed and preyed on the planteaters. And then, 66 million years ago, all these creatures and more disappeared in a global cataclysm.

A semblance of calm eventually returned to the world. No great beasts of any kind crashed their way through the forests, but in the undergrowth little mammals that had been there when the dinosaurs first appeared were still hunting for insects. That scene scarcely altered for hundreds of thousands of years. On a human timescale, such a period seems an eternity. Geologically it was only a moment. But in the history of evolution, it was a phase packed with swift and dazzling invention, for during it the little mammals adapted to fill all the niches vacated by the ruling reptiles and, in doing so, they founded all the great mammalian groups.

Tupaias are only one of those small insect-eating mammals that have survived until today. There are others scattered around the world in odd corners. Many have misleading names that indicate how puzzled people have been about their true nature. In Malaysia, alongside tupaia, lives an unkempt irritable creature with a long nose bristling with whiskers and smelling powerfully of rotten garlic that is known, quite unaccountably, as a moon rat. In Africa, there is the biggest of all, which, because it swims, is called an otter shrew; and a whole group the size of rats which hop, have slender elegant legs and mobile thin trunks and which are known as elephant shrews. In the Caribbean, there are two species of solenodon – one in Cuba, the other on the neighbouring island of Haiti. And Madagascar has a whole group, some striped and hairy, some with spines on their backs, called tenrecs.

But not all are rare or restricted in distribution. That once common inhabitant of the European countryside, the hedgehog, is also an insectivore and is not so dissim-ilar from the rest if, in our mind's eye, we can discount its coat of spines. These are no more than modified hairs and are little indication of true ancestry. And there are also shrews. In many parts of the world they are very abundant indeed, scurrying through the leaf litter in hedgerows and woodlands, seemingly always in a fever of excitement. Although they are only eight centimetres from nose to tail, they are very ferocious, attacking any small creature they encounter – including one another. To sustain themselves, they have to eat great quantities of earthworms and insects every day. Among them is one of the smallest of all mammals, the pygmy shrew, which is so minute that it can squeeze down tunnels no wider than a pencil. Shrews communicate with one another by shrill high-pitched squeaks. They also produce noises of a frequency that is far above the range of our ears, their eyesight is very poor and there is some indication that they use these ultra-sounds as a simple form of echolocation.

Several species of shrew have taken to water in search of their invertebrate food. In Europe, there are two near-relatives called the desmans – one lives in Russia and the other only in the Pyrenees – which use their long mobile noses as snorkels, turning them up so that they project above the surface of the water as their owners swim about busily searching for food.

The shrew group has produced a variant that seeks its prey entirely underground, the mole. Judging from the structure of its paddle-shaped forelegs and powerful shoulders, it seems that the mole's ancestors were once water-living shrews and the mole has simply adapted the same sort of actions for moving along its tunnels. Fur, underground, might be thought to be something of a mechanical handicap, but many moles live in temperate areas and they need fur to keep warm. So it has become very short and without any particular grain so that it points in all directions and the animal can move forwards or backwards along its tight tunnels with equal ease. Eyes are of very little value underground. Even if there were any light to see by, they would easily clog with mud, so they are much reduced in size. Nonetheless, a mole must have some way of finding its prey and it has sense organs at each end. At the front, its main sensor is not its eyes but its nose, which is an organ of both smell and touch, being covered with many sensory bristles. At the rear, it has a short stumpy tail, also covered with bristles, which make it aware of what is happening behind it. The star-nosed mole of America has an additional device, an elegant rosette of fleshy feelers around its nose which it can expand or retract and which give it an extraordinarily sensitive and accurately directed sense of touch.

A mole's tunnels are not simply passageways but traps. Earthworms, beetles, insect larvae, innocently burrowing their way through the soil, may suddenly break into a mole's tunnel. The mole, scurrying along its passages, harvests whatever turns up. Incessantly active, it manages to patrol every stretch of its extensive network at least once every three or four hours and consumes vast numbers of worms each day. On the rare occasions when so many worms collect in the tunnels that even a mole's appetite is sated, it gathers up the surplus, gives each of them a quick bite to immobilise them, and then stores them away, still alive, in an underground larder. Some of these stores have been found with thousands of paralysed worms in them.

A few insectivores specialised early in eating one particular kind of invertebrate, ants and termites. There is no doubt as to what is the best tool for this job – a long, sticky tongue. Many unrelated creatures, taking to this diet, have developed such an organ independently. The numbat, the marsupial anteater in Australia, has one. So has the echidna. Even ant-eating birds, woodpeckers and wrynecks, have developed one that fits inside a special compartment of the skull and in some species extends round the eye socket. But the most extreme version of such a tongue is that evolved by the early placental mammals.

In Africa and Asia, there are eight different kinds of pangolin, medium-sized creatures a metre or so long with short legs and long stout prehensile tails. The biggest of them has a tongue that can extend 40 centimetres beyond its mouth. The sheath that houses it extends right down the front of the animal's chest and is actually connected with its pelvis. The pangolin has lost all its teeth and its lower jaw is reduced to twin slivers of bone. The ants and termites collected by the mucus on the tongue are swallowed and then mashed by the muscular movements of the stomach, which is horny and sometimes contains pebbles to assist in the grinding process.

Without teeth and without any turn of speed, the pangolin has to be well protected. It has an armour of horny scales that overlap like shingles on a roof. At the slightest danger the animal tucks its head into its stomach and wraps itself into a ball with its muscular tail clasped tight around it. In my experience, there is no way in which a pangolin, once rolled, can be forced to unwind. If you want to see what it looks like, the only thing to do is to leave it and let it recover enough confidence to poke its head out nervously and then trundle away.

You might think that it needs protection, not only from predators but from the ants and termites on which it feeds. Its underside is naked except for a few sparse hairs and looks painfully vulnerable. The animal can shut its nostrils and ears with special muscles, but apart from attacks in these hyper-sensitive areas, it seems indifferent to insect bites. It may even welcome them, rather as a bird actively encourages ants to swarm through its feathers and for the same reason. The pangolin sometimes raises its armour and encourages ants to crawl in between its plates and onto its skin, so dealing with parasites that it cannot possibly scratch off itself. Then, according to one story, it shuts its plates with the ants still inside and trots down to the river for a swim so that they are all washed out and its toilet is completed.

South America has its own particular group of insect-eaters which became separated from the rest at a very early stage. Their ancestors were among those placental mammals that, 63 million years ago, migrated down from the north through Panama and mingled with the marsupials. However, the land bridge did not, in this first instance, last for long. After a few million years, it became submerged beneath the sea. So once more the continent was cut off and its animals evolved in isolation. Eventually, contact was re-established and there was a second invasion from the north, as a consequence of which many of the recently evolved South American creatures disappeared.

But not all. The least specialised of the survivors are the armadillos. Like the pangolins, they are protected by armour and it is this that gives them their Spanish-derived name. It consists of a broad shield over the shoulders and another over the pelvis, with a varying number of half-rings over the middle of the back to give a little flexibility.

Armadillos eat insects, other invertebrates, carrion, and any small creatures, like lizards, that they manage to catch. Their standard method of seeking food is to dig. They all have an excellent sense of smell and when they detect something edible in the ground, they suddenly start excavating with manic speed, scattering earth in plumes behind them, their nose jammed into the soil as though they are terrified of losing the scent and frantic to get a mouthful of food as soon as they possibly can. When you watch them, you wonder how they can possibly breathe. In fact they don't. They have the amazing ability to hold their breath for up to six minutes, even while digging. This talent makes credible one of the entertaining stories told about them by the local people in Paraguay. They say that when an armadillo comes to a river, it simply trots down the bank, into water and continues walking unper-turbed along the river bed weighed down by its armour, to emerge dripping on the other side without having even faltered in its stride.

Today, there are some twenty species of armadillo, and once there were a lot more, including a monster with a single-piece domed shell as big as a small car. One such fossil shell has been found which, it seems, was used by early humans as a tent. The biggest surviving species is the giant armadillo, the size of a pig, which lives in the forests of Brazil. It is seldom seen, for it spends most of its days underground in its tunnels, only emerging under cover of darkness. Like all the group, it is very largely insectivorous and consumes great quantities of ants. In Paraguay, the little three-banded armadillo trots about on the tips of its claws like a clockwork toy – this is the one that can roll into a neatly fitting impregnable ball. Down in the pampas of Argentina there are tiny hairy ones that are mole-like and seldom come to the surface. All armadillos have teeth. The giant has about a hundred, which is almost a mammalian record, but they are small, simple and peg-like.

The specialist anteaters of South America, however, like the pangolins of Africa, have lost their teeth entirely. There are three of them. The smallest, the pygmy anteater, lives entirely in trees and exclusively on termites. It is about the size of a squirrel with soft golden fur and curving jaws which form a short tube. A bigger version, the tamandua, is cat-sized, has a prehensile tail and short coarse fur. It too is a tree-dweller but it often comes down to the ground. Out on the open plains, where termite hills stand as thick as tombstones in a graveyard, lives the biggest of the trio, the giant anteater. It is two metres or so long. Its tail is huge, shaggy and flag-like and waves in the breeze as the animal shambles over the savannahs. Its forelegs are bowed, and its claws so long that it has to tuck them inwards and walk on the sides of its feet. With these claws it can tear open termite hills as though they were made of paper. Its toothless jaws form a tube even longer than its forelegs. When it feeds, its huge thong of a tongue flicks in and out of its tiny mouth with great rapidity and runs deep into the galleries of the excavated termite hill.

All anteaters are fairly slow movers. Even a human can outrun the giant. Since they lack teeth as well, they appear fairly defenceless and it seems strange that they should be without the kind of armour which protects both the pangolins and armadillos. But the pygmy anteater and the tamandua favour tree-living ants and termites and spend most of their time up in the branches out of the way of most predators; and the giant anteater is less harmless than might at first appear. If you go to lasso one, it will turn and sweep blindly at you with its forelegs. If it were able to catch you with its huge hooked claws, there would be little chance of your breaking its embrace. There is a tale of the bodies of a jaguar and an anteater being found out on the savannahs, locked together. The anteater had been dreadfully torn by the jaguar's teeth, but its claws were sunk in the jaguar's back and even in death its clutch on its attacker had not been broken.

All these creatures collect crawling insects. But insects also fly. Put up a white screen in a tropical forest at night, illuminate it with a mercury vapour lamp which produces a light particularly attractive to insects, and within a few hours, the screen will be swarming with insects of amazing variety and in extraordinary numbers – huge moths shedding fluff from their wings, mantis with forearms at the ready in bogus piety, beetles moving their legs with the slow inevitability of robots, huge leaping crickets, chafers with bushy antennae, and so many mosquitoes and small flies that they often accumulate in a thick sludge over the lamp and almost blot out the light.

The insects first took to the air some 400 million years ago and had it to themselves until the arrival of the flying reptiles like the pterosaurs, some 200 million years later. Whether the reptiles flew at night we do not know, but it seems unlikely, bearing in mind the reptilian problem of maintaining body temperature. Birds eventually succeeded them, but there is no reason to suppose that there were any more night-flying birds in the past then there are today – which is very few. So a great feast of nocturnal insects awaited any creature that could master the technique of flying in the dark. Yet another variation of the insectivore theme managed to do so.

We have some notion of how the mammals may have managed to get airborne. In Malaysia and the Philippines there lives an animal so odd that zoology has had to give it an order all of its own. This is the colugo. It is about the size of a large rabbit but its entire body, from its neck to the end of its tail, is covered by a softly furred cloak of skin, delicately dappled in grey and cream. When the animal hangs beneath a branch or presses itself against a tree trunk, this skin makes it well-nigh invisible, and when it extends its legs, the cloak becomes a gliding membrane. I was once taken to a patch of woodland in Malaysia where people said there were many of these strange creatures. I searched a likely-looking tree with my binoculars, examining every bump on the trunk and branches with great care. Having convinced myself that it contained nothing, I turned away to scrutinise another, only to see out of the corner of my eye, a huge rectangular shape detach itself and glide silently away. I ran after it, but it landed low down on another trunk over a hundred metres away, and by the time I got there, the animal was quite high and galloping upwards, its two front feet moving forward together and alternating with its hindlegs, its cloak flapping around it like an old dressing gown.

The colugo's gliding technique has several parallels. The marsupial sugar glider planes through the air in just the same way. Two groups of squirrels have also independently acquired the talent. But the colugo has the biggest and most completely enveloping membrane and took to the habit early in mammalian history, for it is certainly a very primitive member of the group and seems to be a direct descendant of an insectivore ancestor. Having perfected a way of life, it has remained unchallenged – and unchanged. There is no link with bats, for its anatomy is different in many fundamental respects, but it is an indication of a stage that some early insectivores may have passed through on their way to achieving flapping flight and becoming those truly accomplished aeronauts, the bats.

That development took place very early, for fossils of fully developed bats have been found that date back to 50 million years ago. They probably expanded to fill one of the empty aerial niches left by the demise of the ptero-saurs, some of which may well have been nocturnal.

The bat's flying membrane stretches not just from the wrist, like the colugos, but along the extended second finger. The other two fingers form struts extending back to the trailing edge. Only the thumb remains free and small. This retains its nail and the bat uses it in its toilet and to help it clamber about its roost. A keel has developed on its chest bone which serves as an attachment for the muscles which beat the wings.

The bats have many of the modifications developed by birds in order to save body weight. The bones in the tail are thinned to mere straws to support the flying membrane or have been lost altogether. Though they have not lost their teeth, their heads are short and often snub-nosed and so avoid being nose-heavy in the air. They had one problem that birds do not face. Their mammalian ancestors had perfected the technique of nourishing their young internally by means of a placenta. The evolutionary process can seldom be put into reverse and no bat has gone back to laying eggs. The female bat must therefore fly with the heavy load of her developing foetus within her. In consequence, it is not surprising to find that bat twins are a rarity, and in almost every case, it is usual for only one young to be born each season. This, in turn, means that in order to ensure the successful perpetuation of its genes, the females must compensate by breeding over a long period, and indeed, bats are, for their size, surprisingly long-lived creatures, some having a life expectancy of around twenty years. Despite the mouse-like appearance of many bats and the inclusion of 'mouse' in their name in other languages such as German and French, bats have their own lineage and are as closely related to us as they are to mice.

Today, all bats fly at night and it is likely that this was always the case, since the birds had already laid claim to the day. To do so, however, the bats had to develop an efficient navigational system. It is based on ultrasounds like those made by the shrews and almost certainly many other insectivores. The bats use them for sonar, an extremely sophisticated method of echolocation using frequencies that lie a long way above the range of the human ear. Most of the sounds we hear have frequencies of around several hundred vibrations a second. Some of us, particularly when we are young, can with difficulty just distinguish sounds with a frequency of 20,000 vibrations a second. Most bats, flying by sonar, use sounds of between 50,000 and 200,000 vibrations a second. Some, like the noctule bat, use lower frequencies, just audible to the keenest human ear. All bats send out these sounds in short bursts, like clicks, twenty or thirty times every second, and their hearing is so acute that from the echo each signal makes, a bat is able to judge the position not only of obstacles around it but of its prey, which is also likely to be flying quite fast. Genetic studies show that echolocation evolved more than once in bats. Even more surprisingly, the same genes that produce echolocation in bats also do the same thing in the only other animal to have this ability – the dolphin.

Most bats wait to receive the echo of one signal before emitting the next. The closer the bat is to an object, the shorter the time taken for the echo to come back, so the bat can increase the number of signals it sends the closer it gets to its prey and thus track its target with increasing accuracy as it closes in for the kill.

Hunting success, however, can mean momentary blindness, for if its mouth is filled by an insect, a bat cannot squeak in the normal way. Some species avoid this problem by squeaking through their noses and have developed a variety of grotesque nasal outgrowths which serve to concentrate the beam of the squeak and act like miniature megaphones. The echoes are picked up by the ears. These too are elaborate, extremely sensitive, and capable, in many cases, of being twisted to detect a signal. So the face of many bats is dominated by sonar equipment – elaborate translucent ears, ribbed with cartilage and laced with an internal tracery of scarlet blood vessels; and on the nose, leaves, spikes and spears to direct the sounds. The combination is often more grotesque than any painted demon in a medieval manuscript. Each species has its own pattern. Why? Probably so that each can produce a unique call. Receptors matched to it alone can then filter out signals from other species.

The system, described in such terms, sounds simple. It seems less so when you encounter it in action. The Gomanton Caves in Borneo contain several million bats, belonging to eight different species. They have lived there for so long that in one chamber their droppings form a huge pyramidal dune spreading across the cave floor and rising 30 metres to the roof. In order to see the bats, we once trudged our way up it. Its surface was covered by a moving, glistening carpet of cockroaches feeding on the guano and a heavy stench of ammonia rose from it. At the top, close to the roof of the chamber, we found the bats roosting in narrow horizontal clefts in the rock. As we shone our torches on them, some detached themselves and flew past us, their wings brushing our faces. Others hung there, twisting their heads with frantic nervousness, to look at us with their black beads of eyes. Beyond we could see thousand upon thousand, packed together as thick and as uniform as heads of grain in a wheat field and swaying in their alarm as though a wind had passed over them. Suddenly, they erupted in panic. Desperate to escape from the confines of their gallery to the main chamber behind us, they came rocketing out in a torrent. By the time we ourselves had retreated from the top of the guano dune the main chamber was a whirlpool of flying bats. Penned by fear of the unfamiliar daylight outside and terrified by our presence inside, they circled round in a vast eddy, filling the air with the sound of their beating skinny wings. We could just detect the lower components of their squeaks, like a cosmic rustling, but their sonar was beyond our ears. The heat from their bodies made the atmosphere, already hot and airless, even more suffocating. We were spattered by their droppings. There were certainly several hundred thousand, flying in a panic round and round beneath the ceiling, as thick as snow flakes driven by a gale. Yet flying at such speeds, they must all have been using their sonar. Why did their calls not interfere with one another, jamming the signals? How could they react with such swiftness that they did not collide? In such places, the dimensions of the problems of sonar navigation seem beyond comprehension.

When evening comes at Gomanton, the bats leave the cave, travelling along regular and restricted paths along the rock ceiling, flying nose to tail and half a dozen or so abreast so that they form one continuous flickering ribbon. They emerge from one corner of the cave mouth at the rate of thousands a minute, a stream of black bodies hurtling out over the forest canopy to begin the night's hunt. The dune of guano at the back of the cave is a testament to their success. A little simple arithmetic shows that every night the colony must catch several tonnes of mosquitoes and other tiny insects.

A few insects have developed systems to protect themselves from bats. In America, there are moths that have the ability to tune in to the frequency of the bat sonar. As soon as they hear a bat approaching, they drop to the ground. Other species go into a spiralling dive which bats find hard to follow. Yet others jam the signal or send back high-frequency sounds that convince the bat that they are inedible and to be avoided.

Not all bats feed on insects. Some have discovered that nectar and pollen are very nutritious, and have refined their flying skills so that they can hover in front of flowers, just like hummingbirds, and gather nectar by probing deep into the blossoms with long thin tongues. Just as a great number of plants have evolved to exploit the services of insects as pollinators, so too some rely on bats. Some cactus, for example, only open their blossoms at night. These are large, robust and pallid, for in the darkness colour is valueless. Their scent, however, is heavy and strong and the petals project well above the armoury of spines on the stem so that the bats are able to visit without damaging their wing membranes.

The biggest of all bats live only on fruit. They are called flying foxes, not only because of their size – and some of them have a wingspan of one and a half metres – but because their coats are reddish brown and their faces are long- nosed and very fox-like. They have large eyes but only small ears and lack any kind of nose-leaf so it is clear that they are not sonar flyers. Fruit bats live not in caves but in the tops of trees in communal roosts tens of thousands strong, hanging like huge black fruit, shrouded in their wings, squabbling noisily among themselves. Occasionally one will stretch a wing and carefully lick the elastic membrane, keeping it meticulously clean and in good flying order. If the day is hot, they may fan themselves with their half-spread wings so that the whole colony seems to shimmer. A sudden noise or a shake of the tree will produce a squall of angry shrieks and hundreds will take off with great flappings of wings, but they will soon resettle. In the evening, they set off in parties to feed. Their silhouette is quite unlike that of birds, for they lack a projecting tail and their flight is very different from the fluttering of insect-hunting bats. Their huge wings beat steadily as long skeins of them keep a level purposeful course across the evening sky. They may travel as far as 70 kilometres in their search for fruit.

Other bats have taken to feeding on meat. Some prey on roosting birds, some take frogs and small lizards, one is reported to feed on other bats. An American species even manages to fish. At dusk, it beats up and down over ponds, lakes, or even the sea. The tail membrane of most bats extends to the ankles. In the fishing bat, it is attached much higher up at the knee, so that the legs are quite free. The bat can therefore trail its feet in the water, keeping the membrane out of the way by folding up its tail. Its toes are large and armed with hook-shaped claws. When it strikes a fish, the bat scoops it up into its mouth and kills it with a powerful crunch of its teeth.

The vampire bat has become very specialised indeed. Its front teeth are modified into two triangular razors. It settles gently on a sleeping mammal, a cow or even a human being. Its saliva contains an anticoagulant, so that its victim's blood, when it appears, will continue to ooze for some time before a clot forms. The vampire then squats beside the wound lapping the blood. They use a very weak form of sonar and it is said that the reason that dogs, whose hearing is also tuned to very high frequencies, are so seldom attacked by them is that they can hear the vampires coming.

All in all there are around 1,200 species of bat. So, taxonomically speaking, one mammal species in every four is a bat. They have made homes for themselves and found sufficient food in all but the very coldest parts of the world. They must be reckoned one of the most successful of the early insectivore variations.

Whales and dolphins, of course, are also warm-blooded, milk-producing mammals and they too have a long ancestry, with fossils dating back to the beginning of the great radiation of the mammals over 50 million years ago.

We have a remarkably accurate set of fossils showing the gradual change of these animals. Their earliest identifiable ancestors were semi-aquatic mammals about the size of a large dog. These were followed by animals with more paddle-like legs about the size of sea-lions.

The major differences between the whales and the early mammals are attributable to adaptations for the swimming life. The forelimbs have become paddles. The rear limbs have been lost altogether, though there are a few small bones buried deep in the whale's body to prove that the whale's ancestors really did, at one time, have back legs. Fur, that hallmark of the mammals, depends for its effect as an insulator on air trapped between the hairs. So it is of little use to a creature that never comes to dry land, and the whales have lost that too, though once again there are relics, a few bristles on the snout to demonstrate that they once had a coat. Insulation, however, is still needed and whales have developed blubber, a thick layer of fat beneath the skin that prevents their body heat from escaping even in the coldest seas.

The mammals' dependency on air for breathing must be considered a real handicap in water, but the whale has minimised the problem by breathing even more efficiently than most land-dwellers. Humans only clear about 15 per cent of the air in their lungs with a normal breath. The whale, in one of its roaring, spouting exhalations, gets rid of about 90 per cent of its spent air. As a result it only has to take a breath at very long intervals. It also has in its muscles a particularly high concentration of a substance called myoglobin, which enables it to store oxygen. It is this constituent that gives whale meat its characteristic dark colour. With the help of these techniques, the fin-back whale, for example, can dive to a depth of 500 metres and swim for forty minutes without drawing breath.

One group of whales has specialised in feeding on tiny shrimp-like crustaceans, krill, which swarm in vast clouds in the sea. Just as teeth are of no value to mammals feeding on ants, so they are no use to those eating krill. So these whales, like anteaters, have lost their teeth. Instead, they have baleen, sheets of horn, feathered at the edges, that hang down like stiff, parallel curtains from the roof of the mouth on either side. The whale takes a huge mouthful of water in the middle of the shoal of krill, half shuts its jaws and then expels the water by pressing its tongue forward so that water is expelled but the krill remains and can be swallowed. Sometimes it gathers the krill by slowly cruising where it is thickest. It can also concentrate a dispersed shoal by diving beneath it and then spiralling up, expelling bubbles as it goes, so that the krill is driven towards, and concentrated in, the centre of the spiral. Then the whale itself, jaws pointing upwards, rises in the centre and gathers them in one gulp.

On such a diet, the baleen whales have grown to an immense size. The blue whale, the biggest of all, grows to over 30 metres long and weighs as much as twenty-five bull elephants. There is a positive advantage to a whale in being large. Maintaining body temperature is easier the bigger you are and the lower the ratio between your volume and your surface area. This phenomenon had affected the dinosaurs but their dimensions were limited by the mechanical strength of bone. Above a certain weight, limbs would simply break. The whales are less hampered. The function of their bones is largely to give rigidity. Support for their bodies comes from the water. Nor does a life spent gently cruising after krill demand great agility. So the baleen whales have developed into the largest living creatures of any kind that have ever lived on earth, four times heavier than the largest-known dinosaur.

The toothed whales feed on different prey. The largest of them, the squid-eating sperm whale, only attains half the size of the blue whale. The smaller ones, dolphins, porpoises and killer whales, hunt both fish and squid and have become extremely fast swimmers, some reputedly being able to reach speeds of over 40 kph.

Moving at such speeds, navigation becomes critically important. Fish are helped by their lateral line system, but mammals lost that far back in their ancestry and the toothed whales have instead a system based on the sounds used by shrews and elaborated by bats, sonar. Dolphins produce the ultrasound with the larynx and an organ in the front of the head, the melon. The frequencies they use are around 200,000 vibrations a second, which is comparable to those used by bats. With its aid, they can not only sense obstacles in their path, but identify, from the quality of the echo, the nature of the objects ahead. In captivity blindfolded dolphins demonstrate that they can, without difficulty, pick out particular shapes of floating rings and will swiftly swim through the water, with blindfolds on their eyes, and exultantly collect on their snout the one shape that they know will bring a reward.

Dolphins produce a great variety of other noises quite apart from ultra-sounds. Some seem to serve to keep a group together when they are travelling at speed, some appear to be warning cries, and some are call-signs so that animals can recognise each other at a distance.

The great whales also have voices. Humpbacks, one of the baleen whales, congregate every spring in Hawaii to give birth to their young and to mate. Some of them also sing. Their song consists of a sequence of yelps, growls, high-pitched squeals and long-drawn-out rumbles. And the whales declaim these songs hour after hour in extended stately recitals. They contain unchanging sequences of notes that have been called themes. Each theme may be repeated over and over again – the number of times varies but the order of the themes in a song is always the same in any one season. Typically, a complete song lasts for about ten minutes, but some have been recorded that continue for half an hour; and whales may sing, repeating their songs, virtually continuously for over twenty-four hours. Each whale has its own characteristic song but it composes it from themes which it shares with the rest of the whale community in Hawaii.

The whales stay in Hawaiian waters for several months, calving, mating and singing. Sometimes they lie on the surface, one immense flipper held vertically in the air. Sometimes they beat the water with it. Occasionally, one will leap clear of the surface, all fifty tonnes in the air, the ridging of its underside plain to see, and fall back with a gigantic surge and thunderous crash. It will breach in this way again and again.

Then, within a few days, the deep blue bays and straits off the Hawaiian islands are empty. The whales have gone. Tens of thousands of humpbacks appear a few weeks later off Alaska, 5000 kilometres away. Tagging has shown that they are the very same animals.

Next spring, they reappear in Hawaii and once more begin to sing. But this time they have new themes in their repertoire and have dropped many of the old ones. Sometimes the songs are so loud that the whole hull of your boat resonates and you can hear ethereal moans and cries coming mysteriously, as from nowhere. If you dive into the peerlessly blue water and swim down, you may, with luck, see the singer hanging in the water below you, a cobalt shape in the sapphire depths. The sound penetrates your body, making the air in your sinuses vibrate in sympathy, as though you were sitting within the widest pipe of the largest cathedral organ, and the whole of your tissues is soaked in sound.

We still do not know exactly why whales sing. We can identify each individual whale by its song, and if we can do so, then surely whales can do the same. Water transmits sound better than air so it may well be that sections of these songs, particularly those low vibrating notes, can be heard by other whales twenty, thirty, even fifty kilometres away, informing them of the whereabouts and activities of the whole whale community.

Anteaters, bats, moles and whales – to such extremes have the descendants of those early protean insectivores gone in the search for their invertebrate food. But there were other sources of nutriment to be tapped as well – plants. Some creatures developed that ate grass and moved out from the forest onto the plains to graze. They were followed by the flesh-eaters, and, in the open, the two interdependent communities evolved, side by side, each advance in hunting efficiency producing responses in defence from the hunted. A second group of creatures found their leaves up in the treetops. Each group must have a chapter of its own; the first, because they are so numerous; and the second, because of our own egocentricity – for those tree-dwellers were our ancestors.

ELEVEN

The Hunters and the Hunted

The forests of today are very much the same, in essence, as those that developed soon after the appearance of the flowering plants, over 50 million years ago. Then as now, there were jungles in Asia, dank rainforests in Africa and South America, and cool verdant woods in Europe. Soft-stemmed herbs and ferns spread across the ground wherever there was enough light, while trees, rising tall, extended their branches into many-tiered ceilings. Everywhere, leaves sprouted; season after season, century after century, they offered an ever-re-newing, inexhaustible supply of food for any animal able to gather and digest them.

Insects claimed their share, gnawing at the wood, scissoring the leaves into fragments. Lizards tore at fronds, and birds, as they acquired a taste for the newly evolving fruit, obliged the plants by distributing their seeds. Small animals, warm-blooded and furry, nibbled away at leaves and seeds. But no large animals dined systematically from this larder of leaves in the wholesale way that the dinosaurs had once done.

Eating plants is no easy business. It demands particular skills and structures just like any other specialist diet. For one thing, vegetable matter is not very nutritious. An animal has to eat great quantities in order to extract enough calories to sustain itself. Some dedicated vegetarians have to spend three-quarters of their waking hours stolidly gathering and chewing leaves and twigs. That process, in itself, may be dangerous, requiring that the creature stand out in the open, exposed to attack. One way for an animal to minimise that risk is to grab as much as possible, as quickly as possible, and run off with it to somewhere safe. This is the strategy of the giant West African rat. It emerges cautiously from its burrow at night, and when it is sure that there is no danger, frantically loads its cheek pouches with anything that looks remotely edible. Seeds, nuts, fruits, roots, occasionally a snail or a beetle, all go in. The pouches are so large that they can hold two hundred or so such morsels. When both sides are crammed full and the rat can barely shut its mouth, its face so swollen that it looks as though it has a dreadful attack of mumps, it scurries back to its burrow. Below ground it empties the whole collection into its larder and begins to sort through it, chewing the edible pieces and putting to one side those objects, such as little bits of wood and small pebbles, that seemed promising at the time but have proved to be disappointments.

Plant-eaters have to have particularly good teeth. Not only do they use them for very long periods but the material they have to deal with is often very tough. Rats, like other rodents – squirrels, mice, beavers, porcupines – cope with this problem by maintaining open roots to their front gnawing teeth, the incisors, so that they continue to grow throughout the animal's life, compensating for wear. They are kept sharp by a simple but very effective self-stropping process. The main body of the rodent incisor is made of dentine, but its front surface is covered by a thick and often brightly coloured layer of enamel which is even harder. The cutting edge of the tooth thus becomes shaped like a chisel. As the top incisors grind over the lower ones the dentine is worn away more quickly and this exposes the blade of enamel at the front, keeping a sharp chisel edge.

Once gnawed, ground and pulped, the food has to be digested. This, too, presents major problems to a vegetarian. Cellulose, the material from which the cell walls of plants are built, is one of the most stable of all organic substances. No digestive juices produced by any mammal make any impression on it. But if the nutritious substances inside the cells are to be released, they have to be broken down in some way. Providing they are not too thick, this can be done mechanically to some extent by chewing. Some bacteria, however, have the rare ability to produce an enzyme that dissolves cellulose, and herbivores maintain cultures of them within their stomachs. The bacteria make a meal of the cellulose and the owner of the stomach can then absorb the cell contents. But even with bacterial help, adequate digestion of a thoroughly vegetarian meal can take a long time.

The rabbit deals with this in a straightforward if somewhat disconcerting way. Its meal of leaves, having been shredded by the incisors, ground with the molars and swallowed, goes down to the stomach where it is attacked by the microorganisms and its own digestive juices. Eventually, it passes down into the gut, is moulded into soft pellets and excreted. This usually happens when the rabbit is resting in its burrow. As soon as the pellets emerge, the rabbit turns round and swallows them. Once more they go down to the stomach and the last vestiges of nourishment are extracted. Only after this second processing are they deposited outside the burrow as the familiar dry pellets and abandoned.

Elephants have particularly acute problems, for they eat, in addition to leaves, a great deal of fibrous twigs and woody material. Apart from their tusks, their only teeth are molars at the back of the mouth, which form massive grinders. As they wear down, they are replaced every few years by new ones erupting from behind and migrating forward along the jaw. There are six sets of replacements in all. These molars pulp and crush with enormous power, but even so, the elephant's food is so woody it requires a very long period of digestion to extract anything of value from it. The elephant's stomach, however, is big enough to provide it. A meal taken by a human being normally passes through the body in about twenty-four hours. An elephant's takes about two and a half days to make the same journey, and for most of that time it is kept stewing in the digestive juices and bacterial broth of the stomach. Much earlier in history, some dinosaurs, eating ferns and cycads, had encountered the same problem and solved it in the same way – by becoming giants.

Elephant dung, even after all this protracted treatment, still contains a great deal of twigs, fibres and seeds that have remained virtually untouched. Some plants that have been stripped by elephants for millennia have reacted to the treatment by coating their seeds with rinds thick enough to withstand a prolonged soaking in the digestive juices. The paradoxical consequence has been that now, unless the rind is weakened by passing through an elephant, the seeds are unable to germinate at all.

The most elaborate apparatus for digesting cellulose is the familiar one used by antelope, deer and buffalo, as well as domestic sheep and cows. They clip grass from their pasture with the lower incisors, pressing it against the tongue or the gums of the upper jaw, which has no teeth in the front. They then swallow it immediately and it goes down to the rumen, a chamber of the stomach which contains a particularly rich brew of bacteria. There it is churned back and forth for several hours, squeezed by a muscular bag, while the bacteria attack the cellulose. Eventually, the mash is brought up the throat, a mouthful at a time, to be chewed in a particularly thorough way by the molars: a ruminant's jaw can move not only up and down but backwards, forwards and sideways. This ruminating can be done, however, at leisure and in safety, when the animal has left the exposed feeding grounds and is relaxing in the shade during the heat of the day. Eventually, the mouthful is swallowed for the second time. It goes past the rumen and on to the stomach proper, which has absorptive walls. Now at last the animal gains some benefit from all its labour.

Leaves have one further shortcoming as food. In temperate parts of the world, many disappear almost entirely for months at a time. The creatures dependent upon them must, therefore, make special preparations as winter approaches. Asiatic sheep turn their food into fat and store it in cushions around the base of their tails. Other species not only feed and fatten themselves as much as they can, but reduce the demands of the next few months to a minimum by hibernating.

The trigger for this reaction is not, as one might suppose, simply a drop in the temperature, for an animal kept in a constantly warm room will nonetheless hibernate at the same time as its fellows in the autumnal chill outside. Instead, it is the animal's body clock, synchronised to the earth's movement since birth and continually recalibrated by changes in day length which provide the signal.

A dormouse in autumn is often almost spherical. It finds a hole, screws up its eyes, tucks its head into its stomach, wraps its soft furry tail around itself and allows the heat of its body to seep away slowly. Its heart beat slows considerably. Its breathing becomes so shallow and infrequent that it is difficult to detect at all. The muscles stiffen and the whole body feels as cold as stone. In this state of suspended animation, the body's fuel demands are so low that the fat store can provide enough to keep all the essential processes ticking over for months. Extreme cold, however, can wake an animal. If it is in danger of being frozen solid, then it stirs and begins to shiver violently, warming itself by burning fuel in its muscles. It may even, in such an emergency, squander some of its remaining reserves of fat by trotting about until the worst of the cold is past and it can go back to sleep again. Normally it is only the ticking of the body clock that brings the dormouse and other winter sleepers out of their holes. Their appetites are now huge and urgent, for during the winter they may have lost as much as half of their body weight. But now starvation is over. The leaves once more are sprouting.

With such methods as these, a great variety of animals nourish themselves on the vegetable foods provided by the forests of the world. Up in the topmost branches, the squirrels scamper along the twigs, collecting bark and shoots, acorns and catkins. Some species have developed furry membranes between their hind- and forelegs so that they can glide between the branches.

Up here, too, live the monkeys. Many species will take a wide variety of food – insects, eggs, nestlings and fruit – but others will only take the leaves of particular trees and have special complicated stomachs to deal with them. Life in their precarious elevated world has led all of them to become marvellously agile, with grasping manipulative hands and quick intelligences. This particular combination of talents eventually led to further developments of such consequence that they must be given a chapter to themselves. But their way is not the only one of winning a leaf-eating life above the ground. One of the first mammals to move up into the branches in South America was the sloth, and it adopted a solution almost exactly the opposite to those of the monkeys.

Today, there are two main kinds of sloth, the two-toed and the three-toed. Of these, the three-toed is considerably the more slothful. It hangs upside down from a branch suspended by hook-like claws at the ends of its long bony arms. It feeds on only one kind of leaf, Cecropia, which happily for the sloth grows in quantity and is easily found. Few predators attack the sloth and nothing competes with it for the Cecropia. Lulled by this security, it has sunk into an existence that is only just short of complete torpor. It spends eighteen out of twenty-four hours soundly asleep. It pays such little attention to its personal hygiene that green algae grow on its coarse hair and communities of a parasitic moth live in the depths of its coat, producing caterpillars which graze on its mouldy hair. Its muscles are such that it is quite incapable of moving at a speed of over a kilometre an hour even over the shortest distances, and the swiftest movement it can make is a sweep of its hooked arm. It is virtually dumb and its hearing is so poor that you can let off a gun within just a few centimetres of it and its only response will be to turn slowly and blink. Even its sense of smell, though it is better than ours, is very much less acute than that of most mammals. And it sleeps and feeds entirely alone.

But it has to have some kind of social life. With such blurred and blunted senses, how does one sloth find another in order to breed? There is one clue. The sloth's digestion works just about as slowly as the rest of its bodily processes and it only defecates and urinates once a week. But most surprisingly, to do so it descends to the ground and it habitually uses the same place. This is the one moment in its life when it is exposed to real danger. A jaguar could easily catch it here. There must be some important reason for it to take such apparently unnecessary risks. Its dung and urine have extremely pungent smells, and the sense of smell is the only one of the sloth's faculties that is not seriously blurred. So a sloth midden is the one place in the forest that another sloth could easily find – and the one place, too, where it stands a chance of meeting another sloth, say once a week or so. A sloth's midden is also its trysting place, for when a female is in oestrus she will descend to the ground every day, leaving a pile of dung that advertises her readiness to mate.

The forest floor is not rich in vegetation. In some areas the shade is so dense that there is nothing but a deep, springy layer of rotting leaves with an occasional fungus pushing up from among them. Elsewhere, if the canopy is thinner, there may be small bushes, a few herbs on the ground and some spindly saplings. In Africa and Asia, such plants provide food for miniature antelope – the mouse deer and the duiker. About the size of dogs, they are extremely shy, but to see one, after long hours of waiting, come silently stepping towards you through the dappled shadows, fastidiously nibbling a carefully chosen leaf, is a revelation of forest life never to be forgotten. Although the duiker have their own evolutionary history, ruminants very like them wandered those forests of 50 million years ago.

In South America, their role is played not by hoofed animals but by rodents, the paca and agouti. They have the same sort of shape and size and similarly solitary habits and dispositions. If anything, they are even more nervous and shy. The slightest suspicion of danger or whiff of an unfamiliar scent and they freeze, staring, panic-stricken, with large lustrous eyes. The snap of a twig will then send them careering headlong through the forest.

Browsing on the taller shrubs and saplings requires greater stature and every forest has a small population of creatures, ranging in size from ponies to horses, that do so. They are so secretive, silent, and few in number that they are hardly ever seen – in Malaya and South America, the tapirs, which are nocturnal; in parts of Southeast Asia, the Sumatran rhinoceros, with a slightly hairy hide, the smallest of all its kind and now sadly exceedingly rare; and in the Congo, the okapi, a short-necked primitive cousin of the giraffe, the largest of these creatures but so shy that it was the last big mammal to be discovered by science and was not seen alive by any European before the beginning of the nineteenth century.

All these ground-living forest dwellers, large and small, are solitary creatures. The reason is not hard to find. The shaded forest floor seldom produces sufficient leaves to sustain a large group in one area for any length of time, and in any case, if several animals are to maintain a relationship, they require some kind of communication. It is not possible to see far in the forest and signalling by sound would attract the attention of hunters. So the mouse deer and the agouti and the tapir live in pairs or by themselves. They maintain territories which they mark with dung or secretions of a gland close to the eye and rely for their defence on concealment, melting away into the undergrowth of a territory that they know well to take refuge in secret hidden retreats.

The hunters that seek them are also solitary. The jaguar stalks the tapir, the leopard pounces on the duiker. A wandering bear will eat most things and will certainly tackle a mouse deer if it gets a chance. The smallest of the hunters – genets, jungle cats, civets and weasels – will pursue rats and mice as well as birds and reptiles.

Of all the hunters, the cats are the most specialised for meat-eating. Their claws are kept sharp by being retracted into sheaths. When they attack, they hook their victim with them and then deliver a piercing bite in the neck that severs the spinal cord and brings a swift death.

The long dagger-like tooth on either side of the mouth, just behind the front teeth, typical of a meat eater, is used to slash open the hide of the prey. The jagged ones farther back in the jaw shear bones. They are all the tools of butchery. None of the dogs or cats can really chew. Most simply bolt their food in gobbets. Flesh is far easier to digest than leaves and twigs and the hunter's stomach needs little help.

These lonely nocturnal duels of ambush and detection, flight and pounce, follow the ancient tactics that were established between the plant-eater and the beast of prey in those very first forests. But some twenty-five million years ago new and very different techniques developed. A change in the world's climate and its vegetation drew these protagonists out of the shadows and into the open. Grasslands appeared.

Grass may look to be a simple, almost primitive, plant, little more than leaves with roots. In fact, it is a highly advanced one, bearing tiny, unobtrusive flowers which rely not on insects to distribute their pollen but on the wind that blows so freely and widely across the open spaces where it grows. It produces horizontal stems running close to the surface or just below it. When fire sweeps across the plains, consuming the old dry leaves, the flames pass quickly so that these stems and the root stocks are unharmed and they produce new sprouts almost immediately. They can do this because grass leaves grow, not from the tip as do those of bushes and trees, but from the base. This is also of enormous benefit to the grazing animals for it means that even though the leaves have been cropped, they will continue to grow unchecked and very soon produce another meal.

The grass itself benefits from the presence of the herds, for they trample and eat the seedlings of bushes or trees that might take root on the plain and that would, were they to grow tall, rob the grass of light and eventually displace it. It seems likely therefore that the spread of the grasslands and the evolution of grazing animals proceeded together, step by step.

The plains attracted not only grazers. With no kind of cover in which to hide, they made tempting targets for beasts of prey as they too moved out of the forests in search of meals. Only the largest of the vegetarians, the elephants and rhinoceros, had nothing to fear. In the forest, their ancestors had needed to move through the trees with ease and silence, and this kept them to a certain size, but there was no such limitation out here and they grew still bigger. Their great bulk, together with their tough skin, put them beyond the power of any carnivore. But for smaller creatures, the plains, so full of food, were also beset with danger.

Some sought safety in burrows. Grasslands are marvellous sites for creatures with a taste for tunnelling. The ground is free from the knottings and intertwinings of tree roots. So here they can construct extended tunnel systems without hindrance and many species have taken spectacular advantage of the opportunity.

One of the most dedicated and specialised of all these burrowers is a bizarre rodent, the naked mole rat of East Africa. It eats not the leaves of grass but its roots, together with odd bulbs and tubers. Mole rats live in families and excavate elaborate underground quarters with special dormitories, nurseries, larders and lavatories. Life spent entirely underground in the warm, dry earth of the African plains has changed them dramatically. They have lost the use of their eyes and shed all their fur. Blind, naked, their sausage-shaped bodies covered with grey wrinkled skin, their appearance is not improved by the most grotesque incisor teeth. These project clear of the head in a bony semicircle in front of the face. They are used not only for feeding but as burrowing tools. Gnawing one's way through earth could clearly be a distasteful business, but the mole rat avoids mouthfuls of soil with a technique used by many other gnawers. It puckers its lips behind those extravagantly protruding teeth and so keeps its mouth tight shut while its teeth busily excavate.

When they dig, they work in teams. The one at the front gnaws with feverish speed, throwing the dislodged soil behind it and straight into the face of the second member of the team. Since it, in any case, is blind, this does not seem to worry it unduly and it simply hurls the soil back between its legs into the face of the next in the queue until at length the last member of the line receives it and throws it vigorously out of the end of the tunnel and on to the surface. A patch of ground colonised by mole rats is studded with conical tips of this waste, with plumes of sand spouting from holes in front of them like miniature volcanoes.

Few, if any, predators are able to make a meal of a mole rat. It can dig faster than any cat or dog and it has no need ever to come to the surface. But those burrowers that eat not the roots of grass but its leaves have to emerge from their holes to feed at some time or other and then they can be in considerable danger. The plains of North America are colonised by rodents the size of small rabbits called, somewhat misleadingly, prairie dogs. They not only graze above ground but do so during the day when coyotes, bobcats, ferrets and hawks are about, all creatures that are only too glad to make a meal of a prairie dog if given the chance. Prairie dogs have accordingly developed defences which depend upon a highly organised social system.

They live in huge concentrations called towns which may contain as many as a thousand animals. Each town is divided up into a number of communities called coteries of about thirty individuals, all of whom know one another well. Many have interconnecting burrows. The coteries always have some members on sentry duty, sitting upright on the mound of excavated earth beside the burrow entrance where they can get the best view of what is going on. If one spots an enemy, it lets out a series of whistling barks. Different kinds of predators elicit different calls so that all know not only that there is danger but what the danger is. The call is repeated by others nearby and so spreads through the town, putting everyone on guard. The inhabitants do not immediately take to flight but take up strategic positions close to their holes. From there, standing on their hindlegs, they stare at the intruder, watching its every move. So as a coyote trots through the town, the alarm spreads from coterie to coterie and the intruder is met with fixed glares from the citizens, who let it come tantalisingly close before they duck into their burrows.

The social life of the prairie dog is not limited to defence. The adults, sitting outside their burrows, proclaim their ownership by giving yet another kind of whistle, accompanied by an engaging little leap in the air. During the breeding season, the coterie members keep very much to themselves and defend their boundaries against any intruders. When this tense time is over, they become more relaxed. Citizens move about the town, wandering into one another's areas. If a stranger approaches a resident, the animals cautiously exchange a rather reserved kiss and then inspect one another's anal glands to see if they are actually acquainted. If they are not, then they separate and the visitor eventually departs. But if they discover that they are members of the same coterie, then they kiss open-mouthed, gently groom one another and often move off to graze side by side.

The prairie dogs' grazing is so intense that many of the plants they favour become eaten out. The animals then move to a different part of their territory and let the old pasture lie fallow for some time to recover. They also culti-vate selectively. They do not like sage, one of the commoner and more robust plants on the plains. If a seedling of one takes root or if there is one growing in a newly colonised patch of territory, they do not simply ignore it but cut it down without eating it. As a result, there is more room for the plants they prefer.

Further south, on the pampas of Argentina, the role of the prairie dog is taken by a guinea pig the size of a spaniel called the viscacha. It, too, lives in dense commu-nities that graze together, but it grazes only at dusk and at dawn. Like many creatures that are active in the twilight, they have prominent recognition marks, broad horizontal black and white stripes across the face. They build cairns over their burrows. If they find any sizeable stone during their excavations they laboriously drag it up to the surface and dump it in the pile on the top. What is more, like good farmers, they enthusiastically do the same with any large object they happen to find in their pastures. So if you drop something on the pampas near a viscacha colony, the place to look for it is not where it may have fallen but on top of the viscacha's monument.

The viscacha is another descendant of that battalion of placental mammals which migrated south from North America across the Panama land bridge when it first formed and which, when it broke, were marooned in South America. Just as the forests were colonised by anteaters, armadillos and unique kinds of monkeys, so the grasslands were invaded by other placentals. Some of them developed into very strange creatures indeed. Two have already been mentioned – the giant anteater and the extinct armadillo with a shell two metres high. There were also many grass- and leaf-eaters. The viscacha is not the only survivor of this group; there are also small rabbit-coloured guinea pigs. But once there were herbivores that grew to a great size. One looked like a camel and stood as tall as an elephant. Another, a relation of the sloth, was about seven metres high and lumbered about on the ground, feeding on bushes and trees.

When the Panama bridge was re-established, creatures from the north again spread south and many of these bizarre forms disappeared. Both the giant camel and the sloth died out. It was therefore a sensation when, at the end of the nineteenth century, it was reported that a German settler in Patagonia, right at the southernmost tip of the continent, had found recent signs of the ground sloths. He had been exploring a cave on his estancia and had found at the back of it, behind an odd line of boulders that seemed to divide the cave into two, a pile of huge bones, pieces of skin covered with shaggy brown hair with curious bony nodules embedded in it and lumps of fresh-looking dung. He hung a piece of the skin on a post to serve as a boundary marker, and there, a few years later, a Swedish traveller noticed it. Eventually specimens reached the Natural History Museum in London, where they were pronounced to be the remains of a ground sloth. They seemed so fresh that some believed that the animals might indeed be still alive. The line of boulders looked quite like the foundations of a man-made wall. Grass stems in the dung had clean edges to them as though they had been cut rather than dragged up by the roots. Perhaps, some suggested, the indigenous people had driven these monsters into the cave and had kept them penned there behind a wall, feeding them with bales of grass like semi-domesticated animals.

For a long time there was neither confirmation nor refutation of these romantic speculations. Now, sadly, they have been dispelled. When you go to the cave, you discover that it is vast and the line of huge boulders at the back, which might seem on a diagram to form the basis of a wall, are almost certainly nothing more than a collapsed part of the ceiling. The atmosphere of the cave is both very dry and extremely cold so the dung owes its fresh look to the fact that it was, in effect, freeze-dried. Today, the bleak country around is sufficiently well travelled for there to be no chance of creatures twice the size of cows to be wandering about unnoticed. Nonetheless, we now know that people reached this part of South America between eight and ten thousand years ago and the ground sloth disappeared from this region at about the same time. So at least some humans may have seen these shambling and marvellous giants, even if they may also have been directly involved in their extinction.

At the time that the sloths were evolving in the south, on the other side of the Panama strait in North America, a different group of grass-eaters was developing on the prairies. Their ancestors were forest-living creatures, not unlike tapirs but the size of mouse deer. Their molar teeth were rounded and suited to forest browsing. On the plains, in order to escape their enemies, they began to run faster and faster. The first forms had four toes on the front legs and three on the hind. The longer the limbs, the better they serve as levers and, properly muscled, the faster they can propel their owners. As time passed these grazers lengthened their legs by rising off the ground onto their toes. Eventually, the side toes dwindled and the animal, an early horse the size of a dog, was running on a single elongated middle toe. The ankle bones thus became placed halfway up its legs, the side toes were reduced to internal vestiges called the splint bones, and the nails thickened to form the protective shock-absorbent hooves.

These changes in the limbs were accompanied by others. The grasses of the plains were becoming tougher to chew. They had started to produce in their leaves tiny sharp crystals of silica which wore teeth badly. So the proto-horses changed their rounded molars into bigger and bigger grinders with hard ridges of dentine in them. One of the problems of the grazing life is that an animal, with its head on the ground for such long periods, cannot keep a good lookout for predators. The higher the eyes can be placed on the head the better. This requirement, together with the necessity to provide room for the enlarged molars, resulted in a considerable elongation of the skull. So the early horses evolved into the forms we know today. They spread across the plains of America and eventually, at a time when the Bering Strait was dry, into Europe. From there they spread south and colonised the plains of Africa. Later, they died out in their original American home and only reappeared there some four hundred years ago when they were shipped across by the Spanish conquistadors. But in Europe and Africa, they flourished as horses, donkeys and zebras.

The zebras share the African plains with other running grazers which, during the same period, had been evolving along lines of their own. They were the descendants of the miniature forest antelopes, so like the mouse deer and duikers. They had already elongated their legs for running within the forest, though in a slightly different way from that of the horses, retaining not one toe on the ground but two. Now, out on the plains, their legs grew even longer and they became the cloven-hoofed grazers – antelope, gazelle and deer. Today they flourish in such numbers that they constitute some of the most spectacular assemblages of wildlife to be seen anywhere in the world.

On the edges of the plains in the open bush, where there is still a little cover to be had, the antelope – dik-dik and duikers – remain very like their forest-dwelling relations, small, browsing on shrubs and living alone or in pairs on territories that they mark and defend. Farther out in the open, where concealment is no longer possible, the antelope seek safety in numbers, gathering together in large herds. They lift their heads regularly from grazing to look around, and with so many sharp eyes and sensitive nostrils on the alert, it is virtually impossible for a hunter to take the herd by surprise. If an attack does eventually come, then the fleeing herd bewilders the hunter with a multiplicity of possible targets. A herd of impala explodes into hundreds of individuals, all running in different directions and leaping spectacularly into the air with soaring bounds three metres high.

Keeping together in such numbers makes great demands on the pasture and the herds have therefore to move regularly over great areas. Wildebeest seem able to detect a shower of rain falling as far away as 50 kilometres and will move off to find it and crop the newly springing grass. But this nomadic habit complicates the social arrangements for breeding which, in the forest, based on a single pair, had been so simple. For some – the impala, springbok and gazelles – territory remains nonetheless the basis of their arrangements. Males and females form separate herds. A few dominant bucks will leave the bachelor herd to establish individual territories for themselves. Each marks the boundary of its land, defends it against other males and tries to attract females into it and mate with them. This however is a very demanding business and most of the bucks who undertake it are exhausted and badly out of condition after three months or so. Eventually, they are forced to yield to stronger, more rested rivals and they go back to join the bachelor herd.

The eland, the largest of the antelopes, and the plains zebra are among the few that have finally broken the bond with territoriality altogether. They form herds in which both sexes are always present and the males settle their problems over females by battling between themselves wherever the herd happens to be.

In order to catch the grazers, the predators on the plain have had to improve greatly their own running techniques. They have not taken to moving on the tips of a reduced number of toes, perhaps because they have always needed their toes, armed with claws, as offensive weapons. Their solution is different. They have effectively lengthened their limbs by making their spine extremely flexible. At full stretch, travelling at high speed, their hind and front legs overlap one another beneath the body just like those of a galloping antelope. The cheetah has a thin elongated body and is said to be the fastest runner on earth, capable of reaching speeds, in bursts, of over 90 kph. But this method is very energy-consuming. Great muscular effort is needed to keep the spine springing back and forth and the cheetah cannot maintain such speeds for more than a minute or so. Either it succeeds in outrunning its prey within a few hundred metres and makes a kill or it has to retire exhausted while the antelope, with their more rigid backs and long-lever legs, continue to gallop off to a safer part of the plains.

Lions are nowhere near as fast as the cheetah. Their top speed is about 80 kph. A wildebeest can do about the same and maintain it for much longer. So lions have had to develop more complicated tactics. Sometimes they rely on stealth, creeping towards their victims, their bodies close to the ground, utilising every bit of cover. Sometimes an individual works by itself. But on occasion, members of a pride will hunt as a team – and they are the only cats that do so. They set off in line abreast. As they approach a group of their prey – antelope, zebra or wildebeest – those lions at the ends of the line move a little quicker so that they encircle the herd. Finally, these break cover, driving the prey towards the lions in the centre of the line. Such tactics often result in several of the team making kills, and a hunt has been watched in which seven wildebeest were brought down.

Hyenas are even slower than lions. The best they can manage is about 65 kph and in consequence their hunting methods have to be even more subtle and dependent on teamwork. The females have separate dens where they rear their pups, but the pack as a whole works together and holds and defends a territory. They have a rich vocabulary of sound and gesture with which they communicate among themselves. They growl and whoop, grunt, yelp and whine and at times produce a most terrifying chorus of orgiastic laughs. In gesture, their tails are particularly eloquent. Normally they are carried pointing down. An erect tail indicates aggression; pointed forward over the back, social excitement; held between the legs tight under the belly, fear. By hunting in well-coordinated teams, they have become so successful that in parts of the African plains, they make the majority of kills and the lions merely use their bigger size to bully their way on to a carcass, the reverse of the popular conception of the relationship between these two species.

Hyenas usually hunt at night. Sometimes they set off in small groups of two or three and then a wildebeest is likely to be their quarry. They test the herds by charging them and then slowing down to watch the fleeing animals closely, as if trying to detect any weakness among individuals. In the end, they appear to select one animal and begin to chase it doggedly, cantering after it, snapping at its heels until it is finally goaded into turning and facing its persecutors. When it does that, it is doomed. While it faces one hyena, the other lunges at its belly, sinks in its teeth and holds on. The wildebeest is now crippled. Soon it is disembowelled and dead.

Zebra are a more difficult prey. To hunt them, the hyenas unite to form a large team. They appear to decide that they will try for zebra even before they start. They assemble at a regularly used meeting ground in the evening, greeting each other with lavish care, smelling one another's mouths, necks and heads, standing head to tail and sniffing and licking genitals. The pack then moves off to hunt. They may stop along the boundary of their territory and refresh its markings with urine. Sometimes they stop and cluster round a patch of ground in a frenzy of excited sniffing. As far as can be seen, there is nothing to distinguish such a place from any other: the importance of the event comes from the activity which reaffirms the bonds between them all. When they are in groups like this, they will trot straight past herds of wildebeest, paying no attention to them. At last they sight a zebra and the hunt begins.

Zebras run in family groups of half a dozen or so, led by the dominant stallion. He it is who is likely to sound the alarm with a braying danger call. As the herd gallops away, he takes up the rear, placing himself between the pursuing hyenas and his mares and foals. The hyenas follow in a crescent behind. The stallion will swerve and attack the pack with powerful kicks and bites and even chase the leading hyena, who may be forced to drop back and allow others to make the running. But eventually one of the pack will get past the stallion and begin to snap at a mare or a foal. As the chase relentlessly continues, one gets a tooth-hold on a leg or the belly or the genitals and the animal is dragged down. While the rest of the terrified herd canters to safety, the hyenas leap on the fallen zebra, howling and whooping, ripping it to pieces. Within a quarter of an hour, the entire carcass – hide, guts and bones, everything except the skull – will have disappeared.

So the speed of the antelopes demanded the guile and teamwork of the hunters. This response came not only from members of the cat and dog family. Other kinds of animals also came out on to the grasslands to hunt. One group of them was particularly slow and poorly armed, so that for them teamwork and communication were even more important. Eventually, they became the most wily, artful and communicative of all the hunters on the plains. To trace their history, we have to return to the forest, for it was there that they had their origins, searching for fruit and tender leaves in the tops of the trees.

TWELVE

A Life in the Trees

I f you want to clamber about in trees, two abilities are extremely useful: a talent for judging distances, and a capacity for holding on to branches. A pair of forward-facing eyes that can both focus on the same object can provide the first; and hands with grasping fingers, the second. About two hundred living species have that pair of physical characteristics. They include monkeys, apes and ourselves, and we, rather egotistically, call the whole group the primates.

There is no doubt that the early shrew-like mammals, which were the ancestors of such diverse creatures as bats, whales and anteaters, also gave rise to the primates. Indeed, the pen-tailed tree shrew, which lives in Southeast Asia is sufficiently close to the primates to have once been classified as one. Now genetic data have shown that this is not the case. Its scientific name is Ptilocercus, 'feather-tail', and it is a relative of the tupaia, which served as a reasonable model for those creatures. Most authorities now agree that the early ancestor of the primate must have looked very like it.

This tree shrew (it is not actually a shrew) does not yet have either of the primate hallmarks. Its forepaws have long, separable digits, but as the thumbs cannot be opposed to the fingers, it has not got a true grasp. Furthermore, each finger ends with a sharp claw, not a flat blunt nail. Its eyes are large and lustrous, but they are placed on the sides of its long snout so that their fields only partly overlap. As its name suggests, it can run along branches like a squirrel. Unlike other tree shrews, it is nocturnal and it has the long whiskers and large eyes and ears that are needed for such a life. Smell is also the basis of its social life. It marks its territory with little drops of urine and with scent from glands in its groin and neck. Its nose, which serves it so well, is very long, with well-developed and extensive passages containing scent receptors. The nose ends with two nostrils that are shaped like inverted commas and surrounded by bare moist skin like the muzzle of a dog. Perhaps most surprisingly as it scurries through the palm trees, Ptilocercus feeds on fermented nectar, consuming quantities of alcohol that would intoxicate a human, yet remains perfectly poised and seemingly uneffected.

All in all, it has to be admitted that Ptilocercus seems at first sight to be a very unlikely creature to be related to a monkey. But there is a whole group of primates that share some of its characteristics and which are unmistakably monkey-like in other ways, and these show how the transformation might have taken place. They are called the prosimians, or 'pre-monkeys'.

Typical of the prosimians is the ring-tailed lemur of Madagascar. It is sometimes called the cat lemur, for it is cat-sized, with soft dove-grey fur, forward-facing lemon-yellow eyes and a long furry tail, handsomely ringed with black and white. One of its commonest calls even sounds like the miaow of a cat. But there the resemblance ends. It is not a hunter but, like many prosimians, largely vegetarian.

Ring-tails spend a lot of time on the ground in troops. Scent plays a very important part in their lives. Their nose is nowhere near as well developed as that of Ptilocercus, but it is still very fox-like in proportion and it too has a moist muzzle with bare skin around the nostrils. They have three kinds of scent glands: one pair on the inside of the wrist which opens through horny spurs; another high up on the chest, close to the armpits; and a third around the genitals. With these, the males and, to a lesser extent, the females, produce a barrage of signals. As the troop roisters through the forest, an animal will come to a particular sapling, smell it carefully, checking doubtless on which individual has been there before, then put its hands on the ground, hoist its rear as high as it can and rubs its genitals several times on the bark. Often, within a minute or so, another individual will come along and repeat the performance. Males also grasp a sapling with both hands and swing their shoulders so that they twist from side to side. Their wrist spurs click against the bark, making deep scratches that are impregnated with their musk.

The male ring-tail uses scent not only as a signature but as a means of offence. When he prepares for battle with a rival, he vigorously folds his arms and rubs his wrists against his armpit glands. Then he brings his tail forward between his hindlegs and in front of his chest and draws it several times between his wrist spurs so that it is loaded with scent. Thus armed, rivals face each other on all fours, lift their haunches high and thrash their splendid tails over their backs with the fur bristling, so that the smell is fanned forwards. Troops meeting on the frontier between their territories may do battle in this way for as long as an hour, hopping and skipping, squealing and yawning, and excitedly marking saplings with their wrist spurs.

The ring-tail also spends a lot of its time in trees. Here, where it behaves in a much more monkey-like way, its primate characteristics show their usefulness. The eyes on the front of its head give it a binocular view. Its hands with their mobile fingers and opposable thumbs grasp branches, and fingers ending not in claws but in short nails that in no way interfere with its grip, are sufficiently dexterous to enable the animal to pluck fruit and leaves from the tips of branches. Although it is quite big, it can leap safely from tree to tree.

The ability to grip is put to good use by infant lemurs. Baby tupaias are deposited in a nest on the ground; their mother only visits them every two days, perhaps to prevent attention being drawn to the vulnerable young. The baby lemur, however, is able to cling to its mother's fur and it does so as soon as it is born. So it travels with her wherever she goes and is provided with parental protection at all times.

Ring-tails have one, occasionally two, babies at a time. Mothers often sit about in groups together, grooming and resting on the forest floor. The young will then scramble happily from one female to another; at times a particularly placid mother may have three or four youngsters clinging to her, while another female may lean over and affectionately lick the lot of them.

The ring-tails' limbs all have grasping digits and are all about the same length, so that when they run on the ground or along a branch, they do so on all fours. There are, however, over twenty different kinds of lemurs in Madagascar and most of them spend nearly all their time in trees. The sifaka, a beautiful creature, a little larger than the ring-tail and with a pure-white fur, has become a specialist in jumping. Its legs are considerably longer than its arms and enable it to leap four or five metres from one tree to another. The price it pays for this spectacular feat is the inability to run on all fours. On the few occasions that it does come down to the ground, the shortness of its arms leaves it with no alternative but to stand upright, and it hops with both feet together, using much the same sort of action as it does when jumping from tree to tree.

Sifakas have scent glands beneath their chins; they mark their territory by rubbing them on an upright branch and then reinforce the effect by dribbling urine over the bark, wriggling their hips and slowly drawing themselves up the branch as they do so.

The most arboreal of all the lemurs – it hardly ever comes down to the ground – is a close relation of the sifaka, the indri. It is the biggest of all living lemurs with a head and body nearly a metre long. It is boldly marked with a variable black and white pattern and its tail is reduced to a tiny stump hidden in its fur. Its legs are even longer in proportion than those of a sifaka, and the big toes are widely separated from the rest and about twice the length, so that each foot resembles a huge caliper with which the animal can grasp even thick trunks. It is the most magnificent jumper of all, launching itself with an explosive straightening of the hindlegs and travelling through the air, torso upright, in soaring bounds which it can repeat again and again so that it seems to bounce its way from trunk to trunk through the forest.

Indris also use scent in marking the trees, though to a much lesser extent than the ring-tail – apparently smell does not play such a major part in their lives. Instead, they have another way of proclaiming their ownership of territory. They sing. Every morning and evening, a family fills its patch of forest with an unearthly wailing chorus. Each individual joins in and draws breath in its own time, so that the sound continues unbroken for minutes on end. When they are alarmed, they lift their heads and trumpet a different hooting call which carries for great distances through the forest.

The indris' use of sound seems a very appropriate way of laying claim to an arboreal territory, but of course it does have one disadvantage. It is extremely indiscreet. It gives away your presence and position to any predator that might be seeking you. Up in the branches, this does not trouble the indri. No natural enemy can reach it there, and so it can sing with impunity.

Although the ring-tail, sifaka, indri and several other Madagascan lemurs are active during the day, their eyes have a reflecting layer behind the retina which increases the ability to see in very dim light. This is a characteristic of animals that move at night and strong evidence that these lemurs were nocturnal until quite recently. Many others of their relatives in Madagascar still are.

The gentle lemur, which is about the size of a rabbit, lives in holes in trees. It sits beside the entrance during the day, peering about myopically. When darkness comes, it becomes a little more lively, clambering around with a comic slow-motion deliberation which it seems unable to shake off, no matter how dire the emergency. The smallest of the group is the mouse lemur, with a snub nose and large appealing eyes, which scampers through the thinnest twigs. The indri has a closely related nocturnal equivalent, the avahi, very similar in appearance and size except that its fur, instead of being black and white, is grey and woolly. Oddest and most specialised of all is the aye-aye. Its body is about the size of that of an otter, it has black shaggy fur, a bushy tail and large membranous ears. One finger on each hand is enormously elongated and seemingly withered, so that it has become a bony articulated probe. With this the aye-aye extracts beetle larvae, its main food, from their holes in rotten wood.

Sixty million years ago, there were no prosimians in Madagascar, but only in Europe and North America. The island had split from Africa about 100 million years earlier, together with what are now India and Antarctica. Then, in an entirely chance event, the ancestor of the lemurs – perhaps just a single, pregnant female – was blown across from Africa to Madagascar, clinging to storm wreckage from the continent. Genetic studies tell us that all modern lemurs stem from that single event. Something similar happened 15 million years ago, when an early mongoose-like animal was similarly cast away, arriving on Madagascar. Today it is represented by 12 species, including the cat-like and highly ferocious fossa. The lemurs were able to succeed because there was no prosimian competition. Elsewhere, for the most part, they lost the competition with the monkeys. But not totally, for all living monkeys, with the single exception of the South American douracouli, are only active during the day. Those prosimians that were nocturnal did not have to cope with a head-on confrontation, and some of them still survive.

In Africa, there are several kinds of bushbaby, very similar to mouse lemurs, as well as the potto and the more lissome angwantibo. These last two parallel the gentle lemurs and, like them, move with a grave deliberation. In Asia, there are two medium-sized nocturnal prosimians – a spindly creature, the slender loris from Sri Lanka, and the rather larger and plumper slow loris. Although all these creatures have quite large eyes, they still signpost their trees with scent and use it for route-finding in the dark. Their marks are made with urine, but since all of these animals are relatively small and live among twigs rather than tree trunks, this poses a problem of placement. A jet of urine might easily miss the intended spot, sprinkle another branch or simply fall uselessly to the ground. So they urinate on their hands and feet, rub them together and then enthusiastically plant pungent handprints throughout their territory.

One more prosimian lives in the forests of Southeast Asia, the tarsier. It is the size and shape of a small bushbaby. It has a long near-naked tail tufted at the end, greatly elongated leaping legs and long-fingered grasping hands. But the briefest glimpse of its face is enough to show that it is a very different creature from the bushbaby. It has gigantic glaring eyes. They are 150 times bigger in proportion to the rest of its body, than our own. Indeed, judged in that way, they are the biggest eyes owned by any mammal anywhere. They bulge from their sockets and are fixed in them, so that the little creature cannot give a sidelong glance – as we can – or look out of the corner of its eye. Instead, if it wants to see something to one side, it has to turn its whole head, a manoeuvre that it performs with the same unsettling ease as an owl and for the same reason, swivelling its head through 180 degrees to look directly backwards over its shoulder blades. In Borneo, the local people believe that it can turn its head, even more amazingly, through a complete circle and conclude that the attachment of head to body is therefore much less secure than in other animals. Being at one time headhunters, they thought that the sight of a tarsier in the forest was a sign that a head would be soon lost – a good omen if you were setting out on a headhunting raid but not so good if you had been planning to remain peaceably inside your longhouse.

As well as these spectacular eyes, the tarsier has paper-thin ears, like those of a bat, that can be twisted and crinkled so as to focus on a particular sound. With these two highly developed sensory organs it hunts at night for insects, small reptiles and even fledgling birds. It rests, usually with its torso upright, clinging to a vertical twig. A beetle, rustling clumsily through the leaves on the forest floor, will quickly attract its attention. The head makes a sudden swivel and a nod downwards. The mobile ears twist forward. The beetle blunders on. Then swiftly, in one leap, the tarsier springs downwards, grabs the beetle with both hands and sinks its teeth into it with an expression of ferocious relish, shutting its huge eyes with each crunch of its jaws.

It marks its territory with urine, but after watching it hunt, it is tempting to believe that vision is just as important to it as the sense of smell. A look at its nose not only confirms this but reveals that the animal is quite distinct from all other prosimians. For one thing, the eyes are so huge that there is little room in the front of the skull for the nose itself, and the internal nasal passages are very much reduced in comparison with, say, a bushbaby's. The nostrils are not comma-shaped nor are they surrounded by bare moist skin, as are the noses of lemurs and other prosimians. In this it resembles monkeys and apes and it is tempting therefore to see the tarsier as representing an ancestral form from which all the higher primates are descended. This is not the case, but genetics do show that the tarsier is a member of the group that leads to our lineage and was one of the first to branch off. It is therefore a close relative of those early primates which, 50 million years ago, spread widely through the world, displacing most of the prosimians and ultimately populating both the Old World and the New with monkeys.

Monkeys differ significantly from all the prosimians, except the tarsier, in that their world is dominated not by smell but by sight. Clearly it is very important for creatures of any size living in trees and, on occasion, jumping between them, to be able to see where they are going. So daylight suits them better than darkness and all monkeys, except for the douracouli, are active at that time. Their eyesight is better than that of prosimians. Not only do they see in depth, they have greatly improved colour perception. With this accuracy of vision they can judge the ripeness of distant fruit and the freshness of leaves. They can detect the presence in the trees of other creatures which, in a monochrome world, might be invisible. And they can use colour in their communications between one another: so monkeys, because their colour vision is so good, have themselves become the most highly coloured of all mammals.

In Africa there lives de Brazza's guenon, which has a white beard, blue spectacles, orange forehead and black cap, the mandrill with a scarlet and blue face, and the vervet monkey, the males of which have startling blue genitals; in China, the snow monkey with a metallic golden coat and an ultramarine face; in the Amazon forests, the uakari with a scarlet naked face. These are among the most spectacularly costumed monkeys, but a great number of other species also possess coloured fur and skin. With these adornments they advertise and threaten, proclaim their species and identify their sex.

They also use sound in a similarly extravagant way, for up in the trees, leaping acrobatically through the branches, they are beyond the reach of any predator except perhaps an eagle and need have little inhibition about revealing their presence. Howler monkeys in South America sit morning and evening, and sing in chorus. Their larynx is extraordinarily large and their throats swell into resonating balloons. The resulting chorus can be heard for several kilometres and is said to be the loudest noise produced by animals of any kind. But all monkeys have a varied repertoire of noises. There is no such thing as a dumb monkey.

The monkeys that reached South America and became isolated there when the isthmus of Panama sank beneath the sea, have developed very much along their own lines. That they are all derived from one common stock is deduced from the number of anatomical features they have in common, including yet another detail of those revealing characteristics, the nostrils. All South American monkeys have flat noses with widely spaced nostrils opening to the side, whereas monkeys in the rest of the world have thin noses with forward- or downward-pointing nostrils.

One South American group, the marmosets and tamarins, still use scent a great deal in communication even though they are active during the day. The males gnaw the bark of a branch and then soak it with urine. But they also have extremely elaborate adornments – moustaches, ear-tufts and wig-like crests – which they flaunt during their social encounters; and they threaten one another with high-pitched twittering calls. Their manner of rearing their young also, like their scent-marking, seems primitive for it is reminiscent of lemurs. The infants readily move from adult to adult and often congregate on a particularly long-suffering and patient father.

Marmosets are the smallest of all true monkeys and seem to have moved from the basic monkey way of life to take up an existence that is more like that of a squirrel than a primate, eating nuts, catching insects and licking sap from bark gnawed by their special forward-pointing incisors. The pygmy marmoset has a body length of only ten centimetres. Since they are so small, they tend to run along branches rather than clamber between them and keep their foothold on the bark with claws. This might also appear to be a direct inheritance from their insectivore ancestors but it seems to be a recent reversion, for the embryonic marmoset begins to develop monkey nails on its fingers and only at a later stage in its development do they change into claws.

The marmosets, however, are exceptional. Most monkeys are very much larger than they. Indeed, the primates, throughout their evolutionary history, show a tendency to increase in size. It is not easy to understand why this should be. Perhaps it is that in disputes between rival males, a bigger animal is likely, from sheer size and muscle and speed, to win the day and so pass on the tendency to grow large to its offspring. But greater weight makes increased demands on those grasping hands, and the South American monkeys have developed a unique way of supplementing them. They have turned their tail into a fifth grasping limb. It is equipped with special muscles so that it can curl and twine, and at the end, its inner surface has lost its hair and developed a ridged skin like that on its fingers. So powerful is it that a spider monkey can hang by its tail while gathering handfuls of fruit with both hands.

African monkeys, for some reason, have never developed their tail in such a way. They use it for other purposes. They extend it horizontally when they run along branches, as an assistance in balancing. When they jump, they swing it in such a way that it has some aerodynamic function, helping an individual to change its trajectory so that to some degree it can control where it lands. Even so, it is difficult to believe that the African monkey's tail is as useful to it as the prehensile tail of its South American cousins. Maybe the failure of the African monkeys to use their tail as a climbing aid has meant that, as they grew larger, they found life in the trees increasingly awkward and insecure and so began to spend more time on the ground. It is certainly a fact that there are no monkeys in the New World that are ground-living, whereas in the Old World there are many.

Down on the ground, the monkey tail seems to have less value. Baboons carry theirs with a droop halfway along the length, almost as though it is broken. Their close relatives, the drill and mandrill, have tails that are reduced to a tiny stump. And the same thing has happened in the macaque family.

The macaque is one of the most successful and versatile of all primates. If you wanted to pick a monkey that was bright, adaptable, versatile, resilient, enterprising, tough and capable of surviving in extreme conditions and taking on all comers, the macaque would win hands down. There are about 23 different species together with many subspecies and between them they stretch halfway around the world, having been stopped only by the Atlantic Ocean at one end of their range and the Pacific at the other. One group lives on Gibraltar, the only non-human primate resident in the wild in Europe. Admittedly, it is questionable how wild they are. During the past three hundred years, the British garrison there has regularly imported more from North Africa every time the colony has become reduced in numbers. They were there before the British came, as long ago as Roman times, and it seems that even then people ferried them across the straits as pets. Nonetheless, it is a tribute to the macaque that it has managed to survive on the Rock, one way or another, for so long.

Another macaque species, the rhesus, is one of the commonest monkeys in India, often living around temples where it is held to be sacred. Farther east still, a species has become an able swimmer, paddling and diving in the mangrove swamps in search of crabs and other crustaceans. In Malaysia, the pig-tailed macaque is trained to climb palm trees and pick coconuts for its human masters; and the most northerly of all monkeys is a macaque, living in Japan, where it has developed a long and shaggy coat to protect it from the rigours of very cold winters.

Nearly all macaques spend a great deal of time on the ground. Their hands and eyes, perfected in response to an arboreal life, have pre-adapted them for success in a terrestrial existence. They also have the advantage of a third faculty which has not yet been mentioned – an enlarged and more complex brain.

This was the necessary accompaniment of the other two developments. The separate manipulation of the fingers required additional control mechanisms. The combination of images from two eyes to produce a single picture required integrating circuits. If monkeys were to use their fingers in grasping and investigating small objects, then there had to be the most accurate coordination between hand and eye and this necessitated connections between the two relevant control areas in the brain. Only one section is less used – that concerned with the sense of smell. In the monkey brain, it can be seen that this part, the olfactory bulbs, has become greatly reduced in size and swamped by a huge expansion of the cerebral cortex, the section of the brain which deals with, among other things, the capacity to learn.

The Japanese macaques provide fascinating evidence of how capable monkeys have become at learning. Several troops of them have been studied by Japanese scientists. One troop lives in the high mountains of northern Japan where in winter the snow lies thick. Observers watched the monkeys extend their range into a part of the forest that none of them had explored before. It contained some hot volcanic springs. The monkeys investigated and found that the warm water could provide a delicious bath. A few tried it. Soon the habit spread. Now all the monkeys there take hot baths every winter. The curiosity that led to this discovery, and the adaptability that allowed the animals to incorporate the new activity into their regular behaviour, is typical of the enterprise of the macaques.

Another group demonstrated it in an even more dramatic way. They live on a small islet, Koshima, in southern Honshu, separated from the mainland by a narrow but turbulent tidal race so that the community is, to a very large extent, a closed one. In 1952 a group of scientists began to study it. The animals, at first, were wild and shy, so in order to entice them out into the open, the investigators began to feed them with sweet potatoes. In 1953 a young 3½-year-old female, whom the observers knew very well and had named Imo, picked up a sweet potato as she had done many hundred times before. As usual, it was covered with earth and sand, but Imo, for some reason, took it down to a pool, dipped it in the water and rubbed off the dirt with her hand. How far this action was a consequence of logical thought it is impossible to say, but the fact was that, having done it once, she made a habit of it.

A month later, one of her young companions began to do the same. Four months later, her mother did so. The habit spread among the members of the group. Some began to use not just freshwater pools but seawater. Perhaps they found the salty taste more pleasant. Today, washing food in the sea is a universal habit, long after scientists had stopped feeding the animals. The only individuals that never learned were those that were already old when Imo made her first experiment. They were too set in their ways to change.

But Imo was not finished with her innovations. The scientists also regularly threw down handfuls of unhusked rice on the beach and trod them into the sand, reasoning that it would take the monkeys so long to pick out the grains there would be plenty of time to observe them. They had reckoned without Imo. She grabbed handfuls of the rice, sand and all, scampered away to a rock pool and threw them into the water. The sand dropped to the bottom but the grain floated and she skimmed it off with her hand. Once again, the habit spread and soon everyone was doing it. Their persistence of this washing habit seems to be related to hygiene – macaques that wash and are fastidious have fewer parasitic worms.

This ability and readiness to learn from your companions results in a community having shared skills and knowledge, shared ways of doing things – in short, a culture. The word, of course, is normally used in the context of human societies, but here, among the macaques of Koshima, we can see the phenomenon beginning in a simple form.

Feeding the Koshima macaques has led to another development. They are tough, aggressive little creatures, with powerful teeth which they do not hesitate to use on one another. They are now so familiar with human beings that they are no longer intimidated by them. When someone arrives with a sack of sweet potatoes, they have no hesitation in trying to snatch pieces. It is hardly practical to hand out roots one at a time, so the researchers simply tip them on the beach and retreat. The macaques fall upon the pile, grabbing a root with one hand, stuffing another in the mouth, and run off, hobbling three-footed. A few, however, do better. They gather up several roots, clutch them to their chests with both arms and then manage to run, standing upright on their hindlegs, across the beach to a defendable place in the rocks. If a daily sack of sweet potatoes were to be a permanent feature of their lives over many generations, it is easy to see that the major share of food would go to those that were genetically endowed with the requisite balance and leg proportions to enable them to perform this trick with ease. These would be better fed, and dominate the group. They would reproduce more successfully and their genes would become widespread in the group. So, over a few thousand years, macaques might become increasingly bipedal. Such a change, indeed, did happen in Africa. To trace its origins, we have to go back some 30 million years.

At that time, one group of lower primates were increasing in size. This brought a change in the way they moved through the trees. Instead of balancing on the top of a branch and running along it, they began to swing along beneath it. Swinging successfully involves physical changes. Over time, arms lengthen, for those with longer arms can reach further; the tail can no longer play any part in balancing so it disappears; and the musculature and the skeleton of the body changes in order to support an abdomen that is no longer slung beneath a horizontal backbone but strapped to a vertical one as to a pillar. Those changes produced the first apes.

Today five main kinds survive: the orang utan and the gibbon in Asia, the gorilla and chimpanzees in Africa; and, of course, ourselves.

The great red-haired ape of Borneo and Sumatra is the heaviest tree-dweller in existence. A male may stand over 1½ metres tall, have arms with a spread of 2½ metres and weigh a massive 200 kilos. The digits on all four limbs have powerful grips, so that the animal is best described as being four-handed, and the ligaments of the hip joints are so long and loose that an orang, particularly when it is young, can stick its legs out at angles that seem, to human eyes, painfully impossible. Plainly, they are excellently adapted for the arboreal life.

At the same time, their size does seem to be something of a handicap to them. Branches break under their weight. Often they are unable to get fruit they relish because it is hanging far out on a branch that would never support them. Moving from tree to tree can also cause problems. There is little difficulty if substantial branches from each tree overlap, but that is not invariably the case. The orang deals with that problem either by reaching out until he can clasp a stout branch, or by rocking the tree that he is in until it bends over far enough for him to scramble across.

Ingenious though these techniques may be, they can hardly be reckoned easy or swift. Indeed, sometimes an old male gets so large that he apparently finds the whole process too exhausting and whenever he wants to travel any distance, he comes down and lumbers across the forest floor. There is also evidence that the arboreal way of life is fraught with danger for the orang. A study of adult skeletons showed, rather pathetically, that 34 per cent had, at one time or another, broken their bones.

The males, as they grow old, develop immense pouches which hang down from the throat like gigantic double chins – not simply fat, but true pouches that can be inflated with air. They extend far down the chest across into the armpits and right over the back to the shoulder blades. Although they may have been used by ancestral orangs as resonators to amplify their voice like howler monkeys, the modern orang does not sing. His most impressive sound is his 'long call', a lengthy sequence of sighs and groans which continues for two or three minutes. To produce it, he partly inflates his throat pouch and the call ends with a number of short bubbling sighs as the pouch deflates. But he makes this call infrequently, and most of his vocalisations consist of grunts, squeaks, hoots, heavy sighs and a sucking noise made through pursed lips. It is a varied repertory but a quiet one that can only be heard fairly close by. The animal more often than not is alone, and during these monologues he gives the impression of a recluse, mumbling and grumbling to himself in an absent-minded way. Males take up this solitary life as soon as they leave their mothers, travelling and eating by themselves and only seeking company when they briefly come together with a female to mate.

Female orang are about half the size of their mates but they too are solitary animals and travel through the forest accompanied only by their young. This preference for solitude may well be connected with their size. Orangs are fruit-eaters, and, being so big, have to find considerable quantities of it every day to sustain themselves. Fruiting trees, however, are uncommon and widely scattered through the forest, at widely varying intervals. Some only bear fruit once every twenty-five years. Others do so almost continuously for about a century but on one branch at a time. Yet others have no regular pattern and are triggered irregularly by a particular change in the weather such as the sudden drop in temperature that precedes a heavy thunderstorm. Even when they do produce fruit, it may only hang on the tree and be edible for a week or so before it becomes over-ripe, falls or is stolen. So the orang have to make long journeys, continually searching, and may well find it more profitable to keep their discoveries to themselves.

The gibbons, also fruit-eaters, of which there are two main kinds and several species, have followed a very different line of development. Increasing size may have been the stimulus that made apes start to swing beneath branches, but the ancestral gibbons subsequently exploited the new style of locomotion to the full by becoming smaller again. In the end they developed into even more accomplished acrobats than any balancing, branch-running monkey. A gibbon in motion in the treetops is one of the most glorious sights the tropical forest has to offer. With a supple grace that is breathtaking, it hurls itself nine or ten metres across space, grabbing an isolated branch and swinging itself off again in another dazzling swoop through the air. The arms that enable it to do this are as long as its legs and torso combined; so long, in fact, that on the rare occasions that it comes down to the ground, they cannot be used as props or crutches, but have to be held above its head out of the way. Its versatile grasping primate hands have also become specialised at the cost of some of their manipulative abilities. Swinging at gibbon speed requires that the hands be used as hooks that can be latched swiftly on to a branch and then detached almost instantaneously. Thumbs get in the way, so they have moved down towards the wrist and become much reduced in size. In consequence, a gibbon cannot pick up small things from the ground with its thumb and forefinger. Instead, it has to cup its hand and gather the object up with a sideways sweep.

Because they are small, there is usually enough fruit on a tree to satiate several of them, so it is practical for them to travel together and they live in tightly knit families. A pair is accompanied by up to four of their offspring of varying ages. Every morning, the family sings in chorus. The male starts with one or two isolated and tentative hoots, others join in, the group launches into an ecstatic song and finally the female takes over with a rising peal that gets faster and faster and higher and higher until it becomes a high trill of a tonal purity that no human soprano could ever challenge. The parallel with the indri of Madagascar is an obvious one. Because of their different ancestral histories, one creature uses its forelimbs as its major propellant, the other its hind. Otherwise, the tropical rainforest in different parts of the world has produced creatures that are remarkably similar – families of singing, vegetarian gymnasts.

The two African apes, in great contrast to their Asian relations, are much more terrestrial in their habits. Gorillas live in central Africa, one form in the forests of the Congo basin, another slightly larger one in the cool sodden moss-forests that cover the flanks of volcanoes on the borders of Rwanda and Zaire. Young gorillas often climb trees, but they do so rather gingerly and without the solemn universal-jointed confidence of orangs. This is hardly surprising. The gorilla foot cannot grasp in the way that an orang's can, so the arms have to provide the main means of hauling up the body. When gorillas descend, they do so feet-first, lowering themselves with their arms, sometimes sliding down, braking by pressing the soles of their feet flat on the trunk and showering moss, creepers and bark all around them.

The big adult males are so huge, weighing up to 275 kilos, that only the stoutest trees can support them. They climb rarely and do not have much reason to, for although the shape of their teeth and the nature of their digestive system suggest that they were once primarily fruit-eaters, like the orang, they now subsist very largely on vegetation that can be reached without climbing, such as nettles, bedstraw creeper and giant celery. Usually, they also sleep on the ground, making a bed among the flattened vegetation on which they have fed.

They live in family groups of a dozen or so, each being led by a great silver-backed patriarch, who has several adult females attached to him. They sit quietly grazing, ripping huge handfuls of stems from the ground with slow, irresistible sweeps of their immense hands, lolling among the dense nettles and celery, sometimes grooming one another. For the most part they sit in silence. Occasionally they exchange quiet grunts or gurgles, and if an individual wanders away from the main group it makes a little belching sound every now and then so that the rest know where it is.

While the adults doze, the young play and wrestle and occasionally rear up on their hindlegs to beat a quick tattoo on their chests, rehearsing the gesture the adults use in display. The silverback leads and protects his group. If he is frightened and angered by intruders he may roar defiance and even charge. A blow from his fist can smash a man's bones. Pestered by a younger rival, who may be trying to lure away one of the females of his group, he will even fight. But the bulk of his days are spent quietly and in peace.

Several groups of gorillas have been studied for many years and, through the patience and understanding of the scientists, have come to accept other people, provided they are properly introduced and behave in a proper fashion. Encountering a gorilla family and being allowed to sit with them is a moving experience. They are in many ways so like us. Their sight and sense of hearing and smell are closely similar to our own, so that they perceive the world in very much the same way as we do. Like us, they live in largely permanent family groups. Their life expectancy is about the same as ours and they move from childhood to maturity and from maturity to senility at very similar ages. We even share the same kind of gestural language and one that you must observe when you are with them. A stare is rude or, put in a less anthropocentric way, threatening – a challenge that invites reprisal. Keeping the head low and the eyes down is a way of expressing submission and friendliness.

The placid disposition of the gorilla is connected with its diet and what it has to do to get it. It lives entirely on vegetation of which there is an infinite supply growing immediately to hand. As it is so big and powerful it has no real enemies and there is no need for it to be particularly nimble in either body or mind.

The other African ape, the chimpanzee, has a very different diet – and temperament. Whereas a gorilla may eat two dozen kinds of leaves and fruit, the chimpanzee samples two hundred or so and, in addition, termites, ants, honey, birds' eggs, birds and even small mammals like monkeys. To do this, it has to be both agile and inquisitive.

Several groups of chimpanzees, living in the forests on the eastern shores of Lake Tanganyika, have been studied by a Japanese team and are now so accustomed to the presence of human beings that you can sit among them for hours at a time. The size of their groups varies, but they are very much bigger than those of the gorilla and may contain as many as fifty animals.

Chimpanzees are adept climbers, sleeping and feeding in trees, but they habitually travel and rest on the ground, even in thick forest. There they move on all fours, their hands knuckle-down and their long stiffly held arms keeping their shoulders high. Even when the group is settled and at ease on the ground, there is constant activity. Youngsters chase one another up trees and play tag and king-of-the-castle. One may practise bed-making, bending over branches in a tree-crown to build a platform, but it will probably tire of it before it is finished and scamper down and do something else.

The sexual bonds between individuals are variable. Some females and some males are monogamous. Other males will mate with many females, and the females themselves, when their hind-quarters inflate into pink fleshy cushions and they become sexually receptive, often court and mate with numerous males. The tie between the young and their mothers is very close. Immediately after birth, the infant clings to its mother's hair with its tiny fists, though at first it is not strong enough to stay there for long without maternal support. It will remain close to its mother, riding on her back like a jockey when the group travels, until it is about five years old. This close dependence, made possible by the baby's grasping hands, has a profound effect on chimpanzee society, for as a result the young learn a great deal from their mother and she is able to keep a close eye on them as they grow up, supervising what they do, pulling them back from danger, showing them with her own example how to behave.

There is a constant interplay between adults in a resting group. New arrivals will greet one another, proffering the back of their outstretched hand to be sniffed and touched with the lips. Elderly males, grey and balding, with bright eyes and wrinkled faces, often sit away from the main activity. They may be as much as forty years old and they often give an expression of short-tempered irascibility. They are treated with considerable respect, the females rushing up to them, smacking their lips and effusively hooting. All of the group, young and old, spend hours grooming one another, carefully sorting through the coarse black hair, scratching the skin with a fingernail to remove a parasite or a scale. So anxious are they to perform this service to one another and so pleasurable do they find it that sometimes a chain of five or six individuals may form, each absorbed in grooming one another. It has become a truly social activity and a gesture of friendship.

One way or another, the group investigates everything around it. A log smelling odd is carefully sniffed and probed with a finger. A leaf may be plucked, scrutinised with the greatest care, explored with the lower lip and gravely handed to others for a similar examination; and then thrown away. The group may visit a termite hill. On the way there, an animal will break off a twig, trim it to a particular size and strip it of its leaves. On arriving at the termite hill, it pokes the twig into one of the holes. When it pulls it out again, it is covered with soldier termites that have gripped it with their jaws in an attempt to defend the nest against the intrusion. The chimp draws the stem through its lips, taking off the insects and eating them with relish. Chimps not only use tools but make them.

The move made so long ago by the early primates from a ground-based scent-dominated often nocturnal existence, to a life in the trees, led to the development of grasping hands, long arms, stereoscopic colour vision and increased brain size. With the aid of these talents, the monkeys and apes have made a great success of their arboreal life. But those of them that subsequently returned to the ground, whether it was because of increasing body size or some other reason, found that these very talents could be deployed in their new situation in a manner that opened up fresh possibilities and led to further changes. The enlarged brain led to an increase in learning and the beginnings of a group culture; the manipulative hand and the coordinated eyes made possible the use and manufacture of tools. The primates that are practising these skills today, however, are in essence repeating a process that another branch of their family started soon after the ancestral apes first appeared in Africa twenty million years ago. It was this branch that eventually stood upright and developed their talents to such a degree that they came to dominate and exploit the world in a way that no animal had ever done before.

THIRTEEN

The Compulsive Communicators

*H*omo sapiens has suddenly become the most numerous of all large animals. Ten thousand years ago, there were perhaps four million individuals in the world, although it is hard to estimate such things accurately. They were ingenious, communicative and resourceful, but they seemed, as a species, to be subject to the same laws and restrictions which govern the numbers of other animals. Two thousand years ago, following the widespread adoption of agriculture, their numbers had risen to three hundred million; and by about 500 years ago, the species had reached 1 billion, and was beginning to overrun the earth. Today, there are over 7.6 billion. By the end of the century, on present trends, there will be nearly 12 billion. These extraordinary creatures have spread to all corners of the earth in an unprecedented way. They live on the ice of the Poles and in the tropical jungles on the equator. They have climbed the highest mountains where oxygen is cripplingly scarce and dived down in special suits to walk on the bed of the sea. Some have even left the planet altogether and visited the moon.

Why did this happen? What power did we acquire that turned us into the most successful of all species? The story starts five million years ago in the great rift valley of East Africa. The grass- and scrub-covered valley bottom was then much as it is today. Some of the creatures that lived there were giant versions of modern species – a pig as big as a cow with tusks a metre long, an immense buffalo, and an elephant standing a third again as tall as the one that is found there now – but others were very like contemporary species such as zebra, rhinoceros and giraffe. There was also a group of apes that had recently become isolated from their relatives. They were just one branch of the descendants of forest-living apes that had been widespread through not only Africa but Europe and Asia about ten million years ago. The first fossils of these plains-living apes were discovered in southern Africa and they were accordingly named Australopithecus, Southern Ape. Now Australopithecus is no longer seen as our direct ancestor, but some kind of cousin. Several more kinds have been discovered in Africa and a great deal of work is going on in an attempt to disentangle their genealogies. Every time a fresh piece of fossil evidence is unearthed, the debates are renewed with great intensity, for all researchers are agreed that these creatures are part of the human family tree. In science's rather cumbersome jargon, they are 'hominins'.

Their fossilised bones are still rare, and scientists are still arguing over where they all fit on our family tree, but enough have been found to give a fairly clear idea of what they were like in life. Their hands and feet resembled those of their tree-climbing ancestors and were very good at grasping things with nails on the digits, not claws. The limbs were not particularly well suited to running and were certainly not nearly as effective for that purpose as those of either the antelopes or the carnivores. Their skulls also show clear signs of their forest-dwelling past. The eyes, as can be judged from the sockets, were well developed. Clearly sight was of great importance for these animals, as it is for all monkeys and apes. By contrast their sense of smell may have been relatively poor, for the skulls have short nasal clefts. The teeth are small and rounded and not well suited to grinding grass or pulping fibrous twigs. Neither do they have shearing blades, like those of a carnivore. On what, therefore, did these creatures feed? They may have grubbed up roots and gathered berries, nuts and fruit, but they also, in spite of the inadequacies of their anatomy, became hunters.

The structure of their hip bones shows that, right at the beginning of their colonisation of the plains, they began to stand upright. The tendency towards a vertical torso was already present among the tree-living primates that used their hands for plucking fruit and leaves. Many of these had also been able to stand up on their hindlegs for short periods when they descended to the ground. On the plains, however, each of the slow steps of moving to a permanent upright posture would have given a slight advantage. Those early hominins were small, defenceless and slow, compared with the predators of the plains, so advance warning of the approach of enemies would have been of the greatest importance and the ability to stand upright and look around might make the difference between life and death. It would also have been of great value in hunting. All the predators on the plain – lions, hunting dogs, hyenas – gather a great deal of information from smell. But for those early hominins, sight was the most important sense, as it had been in the trees. There was more to be gained from getting the head high and looking into the distance than there was from sniffing a patch of dusty grass. The patas monkey, which spends almost all of its time in open grassland, adopts just such tactics, standing up on its hindlegs whenever it is alarmed.

The upright stance is certainly not a way of achieving speed. If anything, it must have slowed down the early hominins. A highly trained human athlete, probably the best two-legged runner there has ever been among primates, can barely maintain a speed of 25 kph for any distance, whereas monkeys, galloping on all fours, can go twice as fast. But bipedalism did bring one further advantage. The hominins had hands with a precise and powerful grip, developed by their ancestors in response to the demands of a tree-climbing life. If they stood upright, these hands could be ready at all times to compensate for the lack of teeth and claws. If the animals were threatened by enemies, they could defend themselves by hurling stones and wielding sticks. Faced with a carcass, they might not be able to open it with their teeth as a lion could do, but they could cut it open using the sharp edge of a stone, held in the hand. They could even take one stone, strike it against another and so shape it. Stones deliberately struck in such a way have facets on them that are quite different from those on stones that have been chipped by rolling in streams or split by frost. They can thus be identified, and many such have been found associated with the skeletons of hominins. The animals had become tool-makers, and carrying tools may well have been an added advantage to becoming bipedal. So hominins claimed a permanent place for themselves in the community of animals on the plains.

This state of affairs lasted for a very long time, probably as much as two million years. Then, one group, which had long split off from its chimpanzee relatives, became particularly well adapted to life on the edge of the forest or on the savannah that came and went with changes in the climate. Their feet became more suited to running, lost their ability to grasp and acquired a slight arch. The hips changed, the joint moved towards the centre of the pelvis to balance the upright torso, and the pelvis itself became more bowl-shaped and broader to provide a base for the strong muscles stretching between pelvis and spine that were needed to hold the belly in its new upright position. The spine developed a slight curve so that the weight of the upper part of the body was better centred. Most importantly, the skull changed. The jaw became smaller and the forehead more domed. The brain of its ancestors had been about the same size as that of a gorilla, around 500 cubic centimetres. Now it was double the size. And the animal grew to a height of over a metre and a half. Science has given this creature a name that reflects its new stance and height – *Homo erectus*, Upright Man.

The *Homo erectus* animals were much more skilled tool-makers than their predecessors. Some of the stones they chipped were carefully shaped with a tapering point at one end and a sharp edge on either side, and were of a size that fitted neatly into the hand. Evidence of one of their successful hunts has been unearthed at Olorgesailie in southwest Kenya. In one small area lie the broken and dismembered skeletons of giant baboons of a species that is now extinct. At least fifty adult animals and a dozen young appear to have been slaughtered here. Among their remains are hundreds of chipped stones and several thousand rough cobbles. All are of rock that does not occur naturally within 30 kilometres of the site. The implications are several. The way the stones have been chipped and shaped establishes that the hunters were *Homo erectus*. The fact that the stones come from a distant site suggests that the hunts were premeditated and that the hunters had armed themselves long before they found their prey. Baboons, even the smaller living species, are very formidable creatures with powerful fanged jaws. Few people today, without firearms, would be prepared to tackle them. The numbers killed at Olorgesailie suggest that such hunts were regular team operations demanding considerable skill. *Homo erectus* was clearly, by now, a very formidable hunter indeed.

With an improved skill in making tools, acquiring the control of fire and probably better communication skills, *Homo erectus* became more and more successful. Their numbers increased and the species began to spread. From what is now Sub-Saharan Africa they moved into the Nile valley and northwards to the eastern shores of the Mediterranean, and thence into northern Europe and Asia. A few crossed over a land bridge that once connected Tunisia, Sicily and Italy. Others travelled eastwards round the Mediterranean and up north through the Balkans. Around 900,000 years ago, a group of five people, including two children, walked across some mudflats in what is now Norfolk. They left a jumbled mass of footprints in the mud, which were miraculously preserved and uncovered a few years ago. Various species of *Homo* had covered virtually the whole of Africa, Europe and Asia. But they appear not to have reached the Americas or Australasia.

Despite this spread across half the planet, Africa remained the home of humanity. Around 600,000 years ago, another type of human gradually moved out of Africa, spreading through the Middle East and up into northern Europe and Asia. One of the first places their remains were discovered was the Neander valley in Germany, and so they are known as Neanderthals. These humans were much stockier than ourselves, strong and with pronounced ridges above their eyes. Once thought to have been brutish and uncultured, the Neanderthals are now widely considered to have had a culture similar to our own – there is evidence that they buried their dead, they made jewellery and they may have left deliberate markings in some of their caves. It is not clear if they could speak as we can, but the number of genetic differences between Neanderthals and ourselves are tiny – only 96 relating to protein-producing genes, and none of them is in any of the genes thought to be involved in language. Back in Africa, our own species, *Homo sapiens* – wise man – emerged, probably in East Africa, spreading rapidly across the continent. The oldest modern human fossils currently known are from Morocco, and date to around 300,000 years ago. As far as we can tell from the fossils, these were people like you and me, with the same intellectual abilities, including speech.

At this point the story becomes confused. Genetic studies show us unequivocally that all modern non-African people can trace their origin back to a slow movement out of Africa that occurred about 70,000 years ago, as humans gradually walked their way across the planet, perhaps using rafts to cross some stretches of water. These were not pioneers or migrants with a clear aim of where they wanted to go, but hunter-gatherers, slowly extending their range as they moved in search of food or shelter. We eventually covered the earth, certainly, but mainly by accident. This Out of Africa movement, as it is known, was probably driven by climate change – we know from genetic studies that around this time the total population of humans collapsed to around 12,000 individuals, probably due to famine or disease. We were very lucky to survive. However, this movement of our species was not the first: archaeology has confirmed that there were identifiably modern humans in China around 100,000 years ago; this is confirmed by similarly dated discoveries in the Middle East. The lack of any inherited connection between these populations and our current species shows that these early expansions must have died out, leaving no genetic trace.

Something even more remarkable happened when we left Africa and encountered the local population, the Neanderthals. We mated with them and exchanged our genes, leaving genetic traces in modern populations. All non-African modern populations carry the trace of those events – if you are not of recent African descent, you will carry some of those genes. Recently, scientists have discovered a hitherto unknown group, known as the Denisovans, after the name of the cave in Siberia where a handful of their remains – a tooth, tiny bits of bone – were found. This offshoot of the Neanderthals also mated with the humans who gradually moved through their territory – peoples from Asia and Australasia carry some Denisovan DNA of various kinds. Back in Africa, the same kind of process was taking place – Africans carry stretches of DNA that are not seen in other populations, and were apparently acquired from completely unknown forms of human. Somewhere in Africa, there are fossils corresponding to the donors of that DNA, remaining to be discovered.

About 40,000 years ago, for reasons we do not understand, the Neanderthals and the Denisovans disappeared.

This may have been due to disease, or to competition with modern humans, or simply the result of living in small, fragmented populations. Whatever the case, we were now without close relatives on the planet, and we continued our expansion, reaching the Americas about 12,000 years ago, moving right down to the southernmost tip of Patagonia. As we moved around the planet, something happened to the large animals that lived in the areas we occupied. They became extinct. From the ground sloths in South America to the mammoths of Siberia and the huge marsupials of Australia, they all disappeared. Although climate change and fragmented populations may have weakened those species of megafauna, it seems clear that, in many cases, our ingenuity and our appetite drove them to extinction. We hunted them to death. The only place this did not happen was Africa, where we had long lived side by side with large mammals.

Humans have an important ability that has helped us to be so successful. Whether or not we shared it with the Neanderthals and the Denisovans is unknown, but we are now the sole possessors of this characteristic. We can imagine what other members of our species are thinking and use gestures to direct their attention or to communicate with them. We may be the only animals that point – a simple gesture that toddlers rapidly acquire, but which requires the ability to imagine what another individual needs to know. Our gestures extend to our face. Human beings have more separate facial muscles than any other animal. They make it possible to move the various elements – lips, cheeks, forehead, eyebrows – in a great variety of ways that no other creature can match.

One of the most important pieces of information it transmits is identity. We take it for granted that all our faces are very different from one another, yet this is a very unusual characteristic among animals. If individuals are to cooperate in an organised team in which each has their own responsibilities, then it is crucial for those taking part to be able to distinguish one from another immediately. Many social animals, such as hyenas and wolves, do this by smell. Human beings' sense of smell, however, was much less informative than their sight, so their identities were proclaimed not by fragrant glandular secretions but by the shape of the face.

Since the features of the face are extremely mobile, they can also convey a great deal of information about changing moods and intentions. We still have little difficulty in understanding expressions of enthusiasm and delight, disgust, anger and amusement. But quite apart from such revelations of emotion, we also send precise messages with our faces – of agreement and dissent, of welcome and summons. Are the gestures we use today arbitrary ones that we have learned from our parents and share with the rest of the community simply because we have the same social background? Or are they deeply embedded in us and an inheritance of our evolutionary past? Some gestures, such as methods of counting or insulting, vary from society to society and are clearly learned. But others appear to be more universal and deep-seated. Did our African ancestors, for example, nod agreement and shake their head in disapproval as most modern populations do? Clues to the answers can come from the gestures used by people from another society who have had no recent contact with our own.

Although there are now virtually no such people, in the last century New Guinea was one of the last places in the world where such people could be found. Even there, very few could be thought to have escaped all influences of the West, for almost every part of the island had been explored. But in the 1960s, one small patch of country in the forested mountains at the headwaters of the Sepik River remained un-entered by outsiders. A pilot, flying over the area, had noticed, in what everyone had assumed to be uninhabited territory, a few huts in clearings. The Australian administrators, who at that time controlled the island, decided to discover who these unknown people were. A patrol was organised, led by a District Commissioner, and I was able to join it. A hundred men from the villages along the river were recruited to carry stores and tents. At the last known village on one of the tributaries, the people, themselves little visited, told us that they knew that somebody lived in the mountains ahead, but no one there had ever met them, knew what language they spoke or even what they called themselves. The river people referred to them as the Biami.

After we had been marching through the mountains for two weeks, drenched by daily rains, living entirely on the food we carried with us, we found footprints. Two people were ahead of us and travelling fast. We followed them. When we broke camp in the mornings, we found their tracks in the forest nearby and knew that they had been sitting watching us the previous evening. That night we left gifts in the forest, but they were not touched. We called greetings in the language of the river people, but we did not know whether or not the Biami could understand it.

In any case, there was no reply. This continued, night after night, until eventually we lost the trail. After three weeks, we had almost given up hope of making contact. Then one morning, we awoke to find seven men standing in the bush within a few metres of our tent. They were very small, and naked but for cane wrapped round their waist with sprigs of green leaves thrust through it in the front and at the back. Some had earrings and necklaces of animal bones. One carried a woven bag full of roots and fruit.

As we scrambled out of our tents, they stood their ground. It was an act of great trust and we tried to demonstrate as quickly and convincingly as possible that our intentions were friendly. The river men spoke to them, but the Biami understood nothing. We had to rely entirely on such gestures as we had in common. And it turned out that there were many of them.

We smiled – and the Biami smiled back. The gesture may seem an odd one as an indication of friendliness, for it draws attention to the teeth, the only natural weapon that a man has. But its essential element is not the teeth but the movement of the lips. In other primates, this is a gesture of appeasement, an indication by a young male chimpanzee, for example, to his dominant senior that he is not challenging authority. In the human species the gesture has become slightly modified by upturning the ends of the mouth and is used to convey welcome and pleasure. We can be sure that this expression has not been entirely learned from our parents and is part of our built-in repertoire of gesture because babies, born deaf and blind, will nonetheless smile when they are picked up to be fed.

We were anxious to extend our relationship with the Biami. We had brought goods for them – beads, salt, knives, cloth – but it seemed condescending and patronising simply to hand them out as gifts. We pointed to their net bag and raised our eyebrows questioningly. The Biami understood immediately and pulled out taro roots and some green bananas. We began to trade. Pointing at an object, touching fingers to indicate numbers, nodding our head in agreement, all these gestures were unambiguous. We all used our eyebrows a great deal. They are the most mobile features of the face. It is possible that they may serve to keep sweat from running in to the eyes, but this does not explain their great mobility. Their main function must surely be as signalling devices. The Biami drew their eyebrows together to express disapproval. When they accompanied this by shaking the head, they made it unequivocally clear that they did not want the beads that we offered. By raising their eyebrows when they examined our knives, they expressed wonder. When I caught the glance of a man standing hesitantly at one side of the group and raised my eyebrows momentarily, at the same time giving a slight backward jerk of my head, the Biami man did the same, a gesture that seemed to be a recognition and a happy acceptance of one another's presence.

This eyebrow flash is used all over the world. It works as well in a Fijian market as in a Japanese store, with nomadic hunters in the Brazilian jungle as in an English pub. Its precise meaning may vary from place to place but that such signals are so widespread and used by such disparate groups suggests very strongly that they are the common inheritance of humanity.

We know a great deal about how those early humans lived, in particular through excavations of sites in southern France and Spain. Along the great limestone valleys of central France such as the Dordogne and in the foothills of the Pyrenees, the cliffs are riddled with caves and almost every one shows some sign of ancient habitation. From the objects that have been found in them, we know a great deal about these people. They used bone needles and sinew to sew clothes of skin and fur. They fished with carefully carved multi-barbed bone harpoons and hunted in the woods with spears tipped with stone blades. Blackened stones show that they had control of fire and they must have treasured it, for it gave them desperately needed warmth in the winter and enabled them to cook meat that their small teeth could not otherwise have chewed. Their teeth, indeed, had become even smaller than those of their ancestors, but their cranium had expanded and was now as big as our own. In short, as far as the skeletons alone are concerned, there is no significant difference between someone who lived in the caves of France 35,000 years ago and ourselves.

The difference between the life of a skin-clad hunter leaving a cave with a spear over the shoulder to hunt mammoth, and a smartly dressed executive driving along a motorway in New York, London or Tokyo, using his or her mobile phone to check their emails, is not due to any further physical development of body or brain during the long period that separates them, but to a completely new evolutionary factor.

We have credited ourselves with several talents to distinguish us from all other animals. Once we thought that we were the only creatures to make and use tools. We now know that this is not so: chimpanzees do so, and so do finches in the Galapagos that cut and trim long thorns to use as pins for extracting grubs from holes in wood. But we are the only creatures to have painted representational pictures and it is this talent which led to developments which ultimately transformed the life of mankind.

Humanity's interest in art goes back deep in our history. In South Africa there are signs that, 100,000 years ago, humans collected ochre from the ground, for what decorative use we do not know, perhaps for painting their bodies, or perhaps for painting on rock walls. No sooner had we arrived in Indonesia, nearly 40,000 years ago, than we painted animals on the cave walls and made outlines of our hands. But the finest flowering of early human artistic ability can be seen in those ancient European caves. The people who lived there ventured deep into the black holes that lead from the back of many of them, finding their way by the feeble flickering light of stone lamps filled with animal fat. There, in some of the most remote parts of the caverns, sometimes in passages and chambers that could only have been reached after hours of crawling, they painted designs on the walls. For pigments they used the red, brown and yellow ochres of iron, and black from charcoal and manganese ore. For brushes, they used their fingers, sticks burred at the end, and sometimes they blew paint on to the rock, probably from the mouth. Sometimes the designs are engraved with a flint tool and there are a few examples of carving in the round, and modelling in clay. Their subjects were almost always the animals that teemed in the area – mammoth, deer, horse, wild cattle, bison and rhinoceros. Often they are superimposed, one on top of the other, giving a striking impression of movement. There are no landscapes and only very rarely human figures. In some caves, the people left a particularly evocative symbol of their visit, the image of their hands made by blowing paint over them so that the outline is left stencilled on the rock. Scattered among the animals, there are abstract designs – parallel lines, squares, grids and rows of dots, curves that some say represent the female genitalia, chevrons that might be arrows. These are the least spectacular of the designs but the most significant for what was to come.

Even now, we do not know why these people painted. Perhaps the designs were part of a ritual – if the chevrons surrounding a great bull represent arrows, then maybe they were drawn to bring success in hunting; if the cattle shown with swollen sides are intended to appear pregnant, then maybe they were made during increase rituals to ensure the fertility of the herds. Maybe their function was less practical and the people painted simply because they enjoyed doing so, taking pleasure in art for art's sake.

Perhaps it is a mistake to seek a single universal explanation. The most ancient of the European paintings is thought to be about 30,000 years old, the youngest maybe 10,000. The interval between these two dates is about six times the length of the entire history of Western civilisation, so there is no more reason to suppose that the same motives lay behind all these paintings than there is to believe that background music saturating a modern hotel serves the same function as a Gregorian chant. But whether they were directed at the gods, at young initiates or appreciative members of the community, they were certainly communications. And they still retain their power to communicate today. Even if we are baffled by their precise meaning, we cannot fail to respond to the perceptiveness and aesthetic sensitivity with which these artists captured the significant outlines of a mammoth, the cocked heads of a herd of antlered deer or the looming bulk of a bison.

Elsewhere in the world it is still possible to discover just what purposes rock painting can have to a hunting people. In Australia, the Aboriginal people still draw designs on rock that are, in many ways, very similar to the prehistoric designs of Europe. They are painted on cliffs and rock shelters, often in parts that are extremely difficult to reach; they are executed in mineral ochres; they are superimposed one on top of another; they include abstract geometrical designs and stencilled handprints; and very often, they represent creatures on which the Aborigines rely for food – barramundi fish, turtles, lizards and kangaroo.

Some of these designs are repainted time and time again, in the belief that by keeping the image of the animals fresh on the rock they will continue to flourish in the surrounding bush. Elsewhere, people paint as an act of worship. Some people in the central desert believe that the world was created by a great spirit snake, the rainbow serpent, whose many-coloured trail appears in the sky after storms. The elders say that it lives in a hole at the base of a long sandstone cliff in the heart of the tribal territory. No one has ever seen the snake itself, though it sometimes leaves the marks of its passing in the sand. Many generations ago, the people painted the snake-god's image on the rock, a huge undulating curve in white ochre, outlined with red. Horseshoe shapes beside it, not unlike some of the geometric designs of prehistory, represent human beings who are descended from the snake. Beside them on the cliff are more symbols, parallel lines and concentric circles, dots and chevrons, that represent the footprints of ancestral animals, carpet snakes and spears.

These designs have been repainted regularly by generations of people. The process of doing so is, in itself, an act of worship, a communion with the snake god creator. Elders regularly visit them to chant the ancient myths and to meditate on their meaning. Relics of the snake, rounded stones engraved with abstract symbols, were kept in clefts in the rocks. The elders took them out reverently, anointed them with red ochre and kangaroo fat, and chanted. Youngsters used to be taken there to be initiated under the image of the snake, to be instructed in the meaning of the symbols, and to witness the re-enactment of the legends in mime and song.

The Aboriginal people are less closely related to the prehistoric cave dwellers of France than we are, but their way of life remains very close to that of early humans. *Homo sapiens* led such an existence, hunting animals and gathering fruits, seeds and roots everywhere in the world for hundreds of thousands of years. Such a life is hazardous and rough. Men, women and children are exposed to the pitiless sifting of an impersonal environment. The slow and the careless are likely to be killed by predators; the weak may starve; the old may fail to survive the torment of a drought. Those whose bodies were, by the chance of genetic variation, better suited to the conditions, had an advantage. They survived and reproduced, handing on that advantage to their children.

So the bodies of men responded to the impress of the world they lived in and made the most recent major physical changes to be incorporated in their genes. In the beginning, we all had dark, protective skins. Dark pigment provides an effective shield. Many indigenous people living in such environments, in Africa, India and Australia, also share another characteristic – they tend to have thin, attenuated bodies. This shape is also a response to their hot dry surroundings. It provides a large area of skin surface in proportion to body weight, a greater expanse over which winds and evaporating sweat can cool the body.

In cold regions, the situation is reversed. The sun's rays, in moderate quantities, are important for health. Without them, the body cannot manufacture vitamin D, so in the north, where the sun is so often hidden, people like the Sámi of Scandinavia have fair skins. The Inuit, living within the Arctic Circle, also have light-coloured skin and, in addition, a physique that is the opposite of the gangling tropical desert-dweller. They are short and squat, the shape with a low surface-to-weight ratio which best retains heat. Their relative lack of facial hair may also be an adaptation to a cold climate, for a beard and moustache, in these conditions, may ice up and become a real impediment.

Since such characteristics as these became fixed in the genes by natural selection, they remain apparent in individuals, generation after generation, no matter where they live, unless the processes similar to those that brought them into existence cause further changes over many thousands of years.

Communities who live by hunting and gathering still exist. Some Aborigines and African bushmen live in deserts. Other groups find all they need from the rainforests in Central Africa and Malaysia. They all live in harmony with the natural world around them, altering it not at all and making do with what it immediately provides. Nowhere are they overwhelmingly numerous. Until recently their life expectancy was relatively short, their birth-rate and the survival of their children curbed by the scarcity of food and the hazards of their lives. Such was our condition for almost all our existence.

With dramatic swiftness, about 8,000 years ago, that began to change. In lands outside the forests and the deserts, the human population began to increase. First in the wetlands of Mesopotamia, rich in game and fish, humans began to grow plants, including lentils and chick-peas to eat, and flax for clothing. Even before this, there is evidence that by use of fire, already sedentary humans in the region had long been deliberately changing the local flora and eating the grains of wild plants. But a change in our fortunes came when we realised that we need not rely on chance encounters with the wild plant. If we forbore to eat all the seeds that were gathered but planted them in a convenient place, such as the rich alluvial soil of Mesopotamia, we would no longer be forced to wander in search of the plant the following summer. We could become farmers, and eventually town-dwellers.

Uruk, in Iraq, was built on what was then the marshy reed-covered delta of the Tigris and Euphrates rivers. Now it is a desert. The town was a complex one. The people planted fields of grain around it and kept herds of goats and sheep. They made pottery, fragments of which still lie all over the site. And in the centre of the town, they constructed an artificial mountain out of baked mud-bricks, held together with layers of plaited reeds. The settled life led by the citizens of Uruk enabled them to make a further crucial advance in humankind's techniques of communication. People who travel perpetually have to keep their material possessions to a minimum. People who live in houses, however, can accumulate all kinds of objects. In the remains of one of the buildings at Uruk a small clay tablet, covered with incisions, was found; thousands more from the same period have since been discovered, in Uruk and in other ancient cities. These are the earliest known pieces of writing. No one yet knows exactly what they mean. They appear to contain records of rations of food and even beer. The shapes seem to be based on the appearance of the objects they represent, but there is no attempt at naturalistic portrayals. The marks are simple diagrams but ones that must have been recognised by the people for whom they were intended.

When those tablets were baked, humans turned their development onto a new course. Now an individual had a means of conveying information to others in a way that was independent of their presence or indeed of their continued existence. People elsewhere and generations unborn could now learn about an individual's successes and failures, his or her insights and strokes of genius. If they had a mind to, they could sift through accumulations of humdrum facts and extract a seed of significance that could lead to wisdom.

Other communities elsewhere, in Central America and in China, made similar innovations as they too developed agriculture. The diagrammatic representations of objects became simplified and took on new meanings. By using them as puns, they could represent sounds. At the eastern end of the Mediterranean, people developed them into a comprehensive system with which they represented every sound they spoke by shapes cut in stone, scored on clay or drawn on paper.

The revolution caused by the sharing of experience and the spread of knowledge had begun. The Chinese, a thousand years ago, gave it further impetus by devising mechanical means of reproducing such marks in great numbers. In Europe, Johann Gutenberg independently, though much later, developed the technique of printing from movable type. Today, our libraries, the descendants of those mud tablets, can be regarded as immense communal brains, memorising far more than any one human brain could hold. More than that, they can be seen as extra-corporeal DNA, adjuncts to our genetic inheritance as important and influential in determining the way we behave as the chromosomes in our tissues are in determining the physical shape of our bodies. It was this accumulated wisdom that eventually enabled us to devise ways of escaping the dictates of the environment. Our knowledge of agricultural techniques and mechanical devices, of medicine and engineering, of mathematics and space travel, all depend on stored experience. Cut off from our libraries and all they represent and marooned on a desert island, any one of us would be quickly reduced to the life of a hunter-gatherer.

Humanity's passion to communicate and to receive communications seems as central to our success as a species as the fin was to the fish or the feather to the birds. We do not limit ourselves to our own acquaintances or even our own generation. Archaeologists labour to decipher clay tablets rescued with painstaking care from Uruk and other ancient cities in the hope that some citizen long ago may have recorded a message of more significance than a boastful genealogy of a chief or a laundry list. In our own cities, dignitaries arrange for messages to be sent to future generations by burying writings in steel cylinders strong enough to survive even a nuclear catastrophe. And scientists, convinced that humanity's most refined and universal language of all is that of mathematics, select a truth that they believe will be recognised through all eternity – a formula for the wavelength of light – and beam it towards other galaxies in the Milky Way to proclaim that here on earth, after nearly four billion years of evolution, a creature has emerged that has for the first time devised its own way of accumulating and transferring experience across generations.

This chapter has been devoted to only one species, ourselves. This may have given the impression that somehow we are the ultimate triumph of evolution, that all these millions of years of development have had no purpose other than to put us on earth. There is no scientific evidence whatsoever to support such a view and no reason to suppose that our stay here will be any more permanent than that of the dinosaurs. Furthermore, there is no reason to suggest that any intelligent animal will arise to replace us.

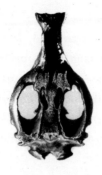

EPILOGUE

Species are not eternal. They come into existence as one kind of animal evolves a particular way to collect food, defend itself and reproduce. But the world around them may change. Rivals may evolve that have new and more efficient ways of feeding. Enemies may appear that are more powerful and dangerous. If such things happen, a species must respond either by abandoning some of its territory and living only in an unchanged part of it, or evolving more effective ways of defending itself and gathering its food. If it does the first it may survive for a very long time indeed. If it does the second, it will, over generations, change and eventually become a new species. If it fails to do either, it becomes extinct.

The human species, however, when it appeared, was exceptional. It was so clever and so dexterous that it did not have to change physically in order to survive on the African plains where it had first evolved. It did not need a physique suited to running at high speed or slashing teeth to catch and kill other animals for food, because it could make weapons to enable it to do so. Nor did it need to change physically to colonise environments colder than the African plains. It could keep warm by wearing clothing made from the skins of the animals it killed.

In due course, these intelligent, inventive humans increased their numbers and spread around the world. No doubt as the end of the Ice Age approached, 40,000 years ago, those in Europe played a part in bringing about the extinctions of the great cold-adapted mammals – giant cattle, sabre-toothed cats and mammoths – that took place as the earth warmed.

The first species of animal that we exterminated in historic times lived on the island of Mauritius in the Indian Ocean. Portuguese sailors landed there at the beginning of the sixteenth century and found great numbers of huge flightless pigeons. The sailors called them dodos, a name that apparently means comic or stupid, for the great lumbering birds were easy prey. The sailors killed great numbers of them and ate them. By all accounts the meat was tough and tasteless, but fresh meat of any kind was what seamen craved. Soon ships of many nationalities were calling regularly at the island to take a share. By the year 1690, the last dodo had been killed.

By this time Europeans had already started to settle on the fertile lands at the southern end of Africa. They found vast plains swarming with herds of a giant antelope called blaubok and a close relative of the zebra which they named quagga, perhaps after the sound of its barking call. The quagga's fore-quarters were striped like its zebra relatives, but its hind quarters were plain brown. The colonists hunted both species for sport and because the wild herds cropped the grass that they wanted for their own domesticated animals. By 1883 both the blaubok and the quagga had been exterminated.

Around the same time as the Dutch started to attack big game in South Africa, British farmers in North America were contending with what some then believed to be the most numerous bird in existence – the passenger pigeon. Its flocks were so vast they darkened the skies, blotting out the sun and taking days to pass by. At night, when they roosted in trees, their weight was so great that branches broke beneath them. Although they were slaughtered in great numbers, they seemed invulnerable. Competitive hunts were organised in which, to qualify for a prize, entrants had to shoot at least thirty thousand birds. The flocks wandered widely in search of food. Sometimes several years would go by without a visit to a particular locality. And then suddenly, people realised that not only had the birds failed to return for several seasons, they had gone for ever. The last survivor, a female called Martha, died in a cage in Cincinnati Zoo in 1914.

Australia, until little more than a century ago, possessed a large native carnivore. It resembled a striped dog but like kangaroos and other marsupials, it carried its young in a pouch. Scientists named it *Thylacinus cynocephalus*, 'pouch-bearer with the head of a wolf', but most Australians called it the Tasmanian tiger and hunted it intensively, for it killed their sheep. Living specimens were sent around the world to be exhibited and wondered at as scientific curiosities. But its numbers dwindled until only one, in Hobart Zoo, remained. It died in 1936. A few may have survived for a little longer in the more remote parts of Tasmania, and optimistic naturalists continued to search for it both there and in the wilder parts of Australia where it had certainly once existed. But no sure signs of one has been found since that time.

These are only a few of the species we have destroyed since the dodo met its end. About ninety more species of birds and thirty-six mammalian species have been exterminated, either directly by hunters or indirectly by the animal predators that we have brought with us as we spread around the world.

It was not until the beginning of the twentieth century that people began to recognise the wholesale damage we are causing. Paradoxically, among the first people to try and halt it, were European big-game hunters. Ever since they had started to explore the wilder parts of Africa, they had competed with one another to collect the biggest horns carried by any particular species. But records established in the past were becoming impossible to match, let alone exceed, no matter how brave and skilful the pursuit and how accurate and deadly the rifles they had. Slowly the hunters realised what was happening. They were exterminating the very animals they so valued. The wilderness was not inexhaustible.

By the 1950s, one species the big-game hunters particularly prized had disappeared almost entirely from the wild.

It was an antelope, the Arabian oryx, which had long, straight horns of incomparable elegance and magnificence. But a few captives still survived in zoos and private collections both in Europe and in the Middle East.

So the hunters became conservationists. All the captive specimens were brought together in Phoenix Zoo in Arizona, where the climatic conditions were not unlike those in the species' native home, and captive breeding began. Because the animals had been collected from different localities in Arabia, the new herd was very varied genetically, so dangers that might have come from in-breeding were minimal. Numbers increased quickly and by 1978, there were enough animals to risk taking some back to Arabia and releasing them. Today there are over a thousand Arabian oryx roaming in the wilderness that was their original home.

The conservation movement had started. The giant panda, then so critically endangered, was adopted as the very emblem of the urgent need for conservation. Attempts by zoos to breed them in captivity, both in Europe and the United States all failed and it seemed that there was a real risk that the species might become totally extinct. But then Chinese zoologists studying captive animals unravelled the species' complex breeding cycle and began breeding them in such numbers that it became possible to return some of them to the bamboo forests that are their original home.

Birds, too, were in urgent need of help. The kakapo, the giant flightless parrot of New Zealand, had once been common throughout the two main islands. But in the nineteenth century, Europeans arrived bringing with them cats and dogs, stoats and rats which found the huge flightless birds easy prey. By the 1980s the species was on the very brink of extinction. But then, in 1987, the survivors, 37 in all, were captured and released on three small offshore islands that were totally free of land-based predators. Its recovery has not been easy, for the bird is a true oddity. It does not, even in the best of circumstances, breed every year and even then, it only lays one or two eggs in a season. So although its numbers are now building, its survival still cannot be taken for granted.

The great whales, the largest animals that have ever existed, easily outweighing the biggest of the dinosaurs, were also imperilled. In the nineteenth century, men had started to hunt them for the sake of the oil-rich blubber that swathed their bodies, protecting them from the cold of the polar seas. At first the difficulties of catching and killing them in the open ocean protected them to some degree, but then in the twentieth century, Norwegian whalers introduced a harpoon gun with an explosive head from which few whales could escape. In the first half of the twentieth century more than 330,000 blue whales were killed in the Southern Ocean alone. Then in 1955, the maritime nations of the world began to see the disaster that was looming and came together. Between them they hammered out an agreement to halt the killing. Now, it seems that the numbers of most, if not all, of the whale species are beginning to recover.

But sadly – and perhaps even more alarmingly – the most widespread and insidious dangers that now face the natural world are not those that humanity has deliberately inflicted, but those they have caused inadvertently. They are not so easily put right. Towards the end of the eighteenth century, people in the north of England began to use coal to generate power and to use it to work machinery of all kinds – to make cloth and other goods, to build railways that could transport people and baggage at unheard-of speeds. The Industrial Revolution that was to spread across the world had started. Smoke and fumes produced by the coal-fired machinery devastated the countryside and choked the atmosphere. Few imagined that such gases could eventually change the chemical character of the atmosphere so extensively that the climate of the entire planet would be affected. But that process had started and is still continuing with increasing speed, bringing devastation with it.

The seas are warming as a consequence of the climatic changes we have caused. They are also seriously polluted by plastic and poisonous waste that we have so carelessly thrown into it, and are now in danger of losing their fertility. It was in the seas that life began over three thousand five hundred million years ago. About six hundred million years later it developed into multicellular organisms, which left evidence of their arrival as faint imprints in the rocks. More and more complex and varied creatures evolved, including some with backbones that gave rise to fish, amphibians, reptiles, birds and mammals.

So the vast number of species that exist today, which we still have not completely catalogued, has come into existence. Its complexity, its vast web of interrelationships and dependencies, is almost beyond comprehension. So complex is it, that we cannot predict with any certainty or accuracy the effects of damage to any one part of it. But that complexity is something we must do our utmost to protect, for it is that which enables it to absorb the worst effects of damage and to heal itself.

Human beings are part of this complex community. We depend on the natural world for every mouthful of food we eat and every lungful of air we breathe. Our health depends on its health. We are now by far the most powerful single species that has ever existed on earth. That power brings great responsibility. It is now up to us to care for the planet and for all the other creatures for whom it is home.

ACKNOWLEDGEMENTS

This book is based on a thirteen-part series with the same title that was filmed in the late 1970s. It came from an idea that I discussed with Christopher Parsons, then the senior producer in the BBC's Natural History Unit in Bristol. He recruited two more producers, Richard Brock and John Sparks, and the four of us together worked out the general shape of the individual programmes.

Once that was done and agreed, we recruited individual natural history cameramen, many of whom had specialist skills, to film the sequences described by the scripts we provided. They in turn worked with scientists who were studying the species we had selected. I learned a great deal from the material they produced, some of which was exactly as scientific texts had predicted and some of it which cast new light on their subjects. Meanwhile, I travelled the world with one or other of the producers and a small crew – Maurice Fisher, Paul Morris and sound recordist Lyndon Bird – filming those sequences in which I had to appear in vision talking to camera. They, too, unknowingly contributed to this book by telling me, unhesitatingly, when the meaning of the words I spoke had become opaque.

Since those days, zoologists have discovered a great deal more about the general structure of the tree of life, largely as a consequence of the discovery of the structure of DNA and its ability to reveal heredity in the 1950s. Happily, this new knowledge has not affected our overall understanding of the structure of the tree of life, but many of the details of its individual branches have changed and become much more defined. In taking account of this latest addition to our knowledge, I am hugely grateful to Professor Matthew Cobb of Manchester University, who has helped me revise the text. His grasp of the detail of the latest zoological research over all branches of zoology is truly extraordinary and he was unfailingly patient in correcting me when my attempts to simplify were leading to error. My debt to him is great indeed.

David Attenborough,
June 2018

INDEX

INDEX

CHAPTER OPENERS

Chapter openers are all from plates and illustrations in Charles Darwin's published works and are reproduced with kind permission from John van Wyhe (ed.), 2002, *The Complete Work of Charles Darwin Online*. (http://darwin-online.org.uk/)

p. 1 Tree frogs (*Hyla* spp.): Plate 19 (detail), Darwin, C. R. (ed.), 1842. *Reptiles Part 5 No. 2 of The Zoology of the Voyage of H.M.S. Beagle*. By Thomas Bell. Edited and superintended by Charles Darwin. London: Smith Elder and Co.

p. 7 Incrustation deposited on tidal rocks: Page 9, Darwin, C. R., 1845. *Journal of Researches into the Natural History and Geology of the Countries Visited During the Voyage of H.M.S. Beagle Round the World, Under the Command of Capt. Fitz Roy, R.N.* 2nd edition. London: John Murray.

p. 35 Shells: Plate IV (detail), Darwin, C. R., 1876. *Geological Observations on the Volcanic Islands and Parts of South America Visited During the Voyage of H.M.S. Beagle.* 2nd edition. London: Smith Elder and Co.

p. 65 Tree fern: Page 10, Darwin, C. R., 1870. *Rejseiagttagelser (1835–6) af C. Darwin. (Tahiti. – Ny-Seland. – Ny-Holland. – Van Diemens Land. – Killing-Øerne.).* 1st edition. Copenhagen: Gad.

p. 92 *Chlorocoelus Tanana* (from Bates): Page 355, Darwin, C. R., 1871. *The Descent of Man, and Selection in Relation to Sex.* 1st edn. London: John Murray.

p. 117 Galapagos gurnard (*Prionotus miles*): Plate 6, Darwin, C. R. (ed.), 1840. *Fish Part 4 No. 1 of The Zoology of the Voyage of H.M.S. Beagle.* By Leonard Jenyns. Edited and superintended by Charles Darwin. London: Smith Elder and Co.

p. 144 *Uperodon ornatum*: Plate 20 (detail), Darwin, C. R. (ed.), 1843. *Reptiles Part 5 No. 2 of The Zoology of the Voyage of H.M.S. Beagle.* By Thomas Bell. Edited and superintended by Charles Darwin. London: Smith Elder and Co.

p. 168 Bibron's Tree Iguana (*Proctotretus bibronii*): Plate 3 (detail), Darwin, C. R. (ed.), 1842. *Reptiles Part 5 No. 1 of The Zoology of the Voyage of H.M.S. Beagle.* By Thomas Bell. Edited and superintended by Charles Darwin. London: Smith Elder and Co.

p. 192 Peacock feather: Fig. 53, Darwin, C. R., 1871. *The Descent of Man, and Selection in Relation to Sex.* Volume 2. 1st edn, London: John Murray.

p. 225 Platypus (*Ornithorhynchus paradoxus*): Page 528, Darwin, C. R., 1890. *Journal of Researches into the Natural History and Geology of the Various Countries Visited by H.M.S. Beagle etc.* London: Thomas Nelson.

p. 252 Vampire bat (*Desmodus rotundus*): Page 37, Darwin, C. R. 1890. *Journal of Researches into the Natural History and Geology of the Various Countries Visited by H.M.S. Beagle etc.* London: Thomas Nelson.

p. 277 Ethiopian warthog (*Phacochoerus aethiopicus*): Fig. 65, Darwin, C. R. 1871. *The Descent of Man, and Selection in Relation to Sex.* Volume 1. 1st Edn. London: John Murray.

p. 302 Diana monkey (*Cercopithecus diana*): Fig. 76, Darwin, C. R. 1871. *The Descent of Man, and Selection in Relation to Sex.* Volume 2. 1st Edn. London: John Murray.

p. 331 Fuegia Basket, 1833: FitzRoy, R. 1839. Page 324, *Proceedings of the Second Expedition, 1831–36, Under the Command of Captain Robert Fitz-Roy, R.N.* London: Henry Colburn.

p. 356 Top view of the skull of a *Toxodon* sp. (extinct): Fig. III, Darwin, C. R. (ed.). 1838. *Fossil Mammalia Part 1 No. 1 of The Zoology of the Voyage of H.M.S. Beagle.* By Richard Owen. Edited and superintended by Charles Darwin. London: Smith Elder and Co.